M-libraries 2

A virtual library in
everyone's pocket

GW00535686

M-libraries 2

A virtual library
in everyone's pocket

edited by
Mohamed Ally and **Gill Needham**

facet publishing

Published by Facet Publishing,
7 Ridgmount Street, London WC1E 7AE
www.facetpublishing.co.uk

Facet Publishing is wholly owned by CILIP: the Chartered Institute of Library and
Information Professionals.

British Library Cataloguing in Publication Data
A catalogue record for this book is available from the British Library.

ISBN 978-1-85604-696-1

First published 2010

Mixed Sources
Product group from well-managed
forests and other controlled sources
www.fsc.org Cert no. SA-COC-1565
© 1996 Forest Stewardship Council

FSC

Text printed on FSC accredited material.

Typeset from author's files by Facet Publishing in 12/14 American Garamond and Nimbus
Sans.
Printed and made in Great Britain by MPG Books Group, UK.

Contents

Acknowledgements

Without the Second International M-Libraries Conference this book would not exist, and so we would like to thank everyone who contributed to its success: the universities responsible for the conference (University of British Columbia, Athabasca University, Thompson Rivers and the Open University (UK)); the International Organizing Committee, Programme Committee and Local Organizing Committee; all our excellent speakers, session chairs, delegates, helpers and commercial sponsors. We would like to give particular thanks to our colleagues Leonora Crema, Nicky Whitsed, Nancy Levesque and Steve Shafer for their brilliance in pulling the conference together, and a special mention to Leonora and her team at UBC for hosting the event with such efficiency and style.

Our thanks go to all the authors who have contributed to the book, not only for their inspired content but for their co-operation in meeting deadlines, and to all our colleagues who have helped to put the book together. Particular thanks go to Jill Gravestock and to Jeannette Stanley.

Mohamed Ally
Gill Needham

Contributors

Dr Mohamed Ally is Director and Professor in the Centre for Distance Education at Athabasca University, Canada. He is currently conducting research in the areas of mobile technology in learning and training, mobile library, e-learning and distance education. He recently presented and published papers on the use of emerging technology to bridge the learning divide, mobile learning and mobile library. He has published papers and chapters in books, journals and encyclopedias. He recently edited a book entitled *Mobile learning: transforming the delivery of education and training* (2009) and previously co-edited, *M-libraries: libraries on the move to provide virtual access* (Facet Publishing, 2008), the predecessor to this title.

Jose Luis Andrade was named President, Swets North America, in September 2007. He is responsible for managing sales, marketing, customer service, finance and operations activities in the United States and Canada. Most recently, he has also acquired all commercial responsibility for Swets Latin America. He was appointed to his current role from his position as Swets General Manager for Latin America, which he held from 2004 to 2007. In this role, he completely restructured the sales operations in the region. He designed and executed a new marketing strategy, which achieved tremendous growth, placing Swets in the top market position within the Caribbean and Latin America. Previously, Jose Luis held numerous executive management roles at several multinational companies, including Exactus Corporation, eshare communications, inc., and Bentley Systems Inc., where he successfully developed and grew the existing markets. He holds a BS degree in Industrial Engineering from Ibero-American University, and completed the Harvard Business School Executive Management course. Jose Luis is a member of the Board of Directors for the Friends of the National Library of Medicine, Special Libraries Association and American Libraries Association. In addition, he regularly contributes presentations and speeches to various publications and related industry forums.

Veronica Arellano is a former reference and instruction librarian at the University of Houston M.D. Anderson Library. She received her MS in Library Science from the University of North Texas in 2003 and was a 2009 ALA Emerging Leader. A public services librarian at heart, Veronica blogs about all things library at http://freelancelibrarian.wordpress.com.

Alison Armstrong is the Director of Information and Instruction Services at the University of Vermont. Her work has focused on undergraduate library services related to teaching, learning and technology. She earned an MA in Library Studies and a BA in English from the University of Wisconsin, Madison.

Parveen Babbar is Assistant Librarian in Indira Gandhi National Open University. He has been associated with University of Delhi, India and has more than eight years of professional experience. He has master's degrees in MLISc, MBA, MCA and MPhil in Library and Information Science, and was Gold Medallist in MLISc from Delhi University. He has taken on many projects of Library Automation and Digitization, such as CALPI Library Setup for Swiss Agency. He is a member of Special Libraries Association and presently holds the position of webmaster in the Board of the Asian Chapter, SLA. He has presented many papers in national and international conferences and published about 16 papers in reputed journals.

Ken Banks, founder of kiwanja.net, specializes in the application of mobile technology for positive social and environmental change in the developing world. He combines over 22 years in IT with more than 16 years' experience living and working in Africa, including Kenya, Nigeria (where he ran a primate sanctuary), South Africa, Mozambique, Cameroon,

Zambia, Uganda and Zimbabwe. In 1999 he graduated from Sussex University with honours in Social Anthropology with Development Studies. His vision is to empower others to create social change, which he does by developing and providing tools to mostly grassroots organizations who seek to make better use of technology in their work. In 2007 he made headline news on the BBC when his text messaging application FrontlineSMS was used to help monitor the Nigerian presidential elections. Since its launch the software has been successfully implemented in over forty countries, including Afghanistan, Indonesia, Zimbabwe, the Philippines and Pakistan. Ken has been interviewed by many newspapers and journals, and he has sat on panels at a number of events including the Aspen Institute and the Clinton Global Initiative. He also sits on Vodafone's 'Social Impact of Mobiles' emerging markets advisory panel. He has written about his work, and the wider role of reporting on mobile technology for a number of publications including *Didactics World*, BBC News, *Boston Review*, *Vodafone Receiver* and Stanford and Harvard university magazines, and has a regular online column in *PC World*. He has also appeared as a television guest, discussing issues on ICT and Africa, and has acted as an official judge for the Global Mobile Awards, the Mobile Messaging Awards and his own nGOmobile initiative. He was also a regional judge for the 2008 Adjudication Panel for the African ICT Achievers Awards Programme.

Eugene Barsky is a Science and Engineering Librarian at the University of British Columbia. Eugene holds a MLIS degree from UBC and previously worked in medical librarianship. He is the winner of the 2007 Canadian Health Library Association 'Emerging Leader' Award and the 2007 Partnership Award from the Canadian Physiotherapy Association.

Miranda Bennett is Program Director for Collections at the University of Houston Libraries, where she has worked since 2005. Her interests in mobile libraries focus on mobile access to library resources and the evolution of e-readers. She is active in several professional organizations, including the Association for Library Collections and Technical Services and the Association of College and Research Libraries.

Dr Seema Chandhok is Deputy Librarian at Indira Gandhi National Open University (IGNOU), New Delhi, India. She has more than 23 years' professional experience, including 18 years as Assistant Librarian in IGNOU. She has master's degrees in Public Administration, Distance Education, Library and Information Science and a PhD in Library and Information Science. She has several publications in professional journals and conference proceedings and has also published indexes to articles published in journals.

Phil Cheeseman has been working at Roehampton University for four years. After heading eLearning Services, he was involved in merging the eLearning Services and the Academic Liaison Librarian teams, and is now Head of Academic Liaison Services. His team provides professional support for teaching and learning throughout the University and is responsible for managing and supporting use of the University's virtual learning environment and other learning technologies. This newly formed team has brought together liaison librarians and e-learning advisers in a combined unit, and provides exciting opportunities for developing and promoting use of library services, particularly those that have a technological emphasis. Prior to his time at Roehampton, Phil was a college lecturer for 11 years, teaching on a broad range of programmes from biology to teacher training. It was in this role that he developed an interest in the use of technology to support learning.

Scott Collard is the Social Sciences Collections Coordinator and Graduate Student Services Coordinator at New York University. He chairs VRSS, which administers the library's virtual reference services, and is the liaison to the Education and Linguistics programmes. Scott received his MSLIS from the University of Illinois, Urbana-Champaign, and an MA in General Studies in the Humanities (Early American History and Literature) from the University of Chicago. Scott is chair of the Membership and Orientation Committee of the Educational and Behavioral Sciences Section of the Association of College and Research Libraries. His research interests focus on the intersection between technology and public services, user-centred design and customizability.

Karen A. Coombs is the Web Application Specialist for LISHost. Prior to LISHost, Karen served as the Web Services Librarian at the University of Houston. She is a regular presenter at library technology conferences and is the author of a book, *Library Blogging* and numerous articles.

Robin Dasler is the Science and Mathematics Librarian at the University of Houston. She is responsible for liaising with the departments of Physics, Math and Geosciences, as well as the College of Technology. In addition to her liaison duties, she is the co-founder of the University of Houston Libraries' Marketing Committee, whose efforts earned the 2009 Texas Library Association Branding Iron Award. Robin is also the co-chair of the Association of College and Research Libraries Science and Technology Section Membership and Recruitment Committee.

Colin Elliott is the Digitization Coordinator for Athabasca University. He coordinates the digitization efforts of the AU Library as well as various mobile learning projects. Colin has been involved with AU's Digitization Portal, Mobile ESL Project, Workplace English and Mobile French projects. He has presented at both domestic and international conferences, including the First and Second M-Libraries Conferences. He co-wrote a chapter in *M-Libraries: libraries on the move to provide virtual access.*

Elizabeth C. Reade Fong is Deputy University Librarian (Customer Services) at the University of the South Pacific, where she has worked since 1978. She is responsible for customer services, human resource management and quality assurance and plays an important role in the implementation of the Library's Strategic Plan. She has a BA in Sociology and Management from the University of the South Pacific and an MA in Information Science from the University of North London. Elizabeth is a Fijian citizen and has been a member of the Fiji Library Association since 1978, where she has held a number of executive positions, including that of President and Convenor for National Library Week celebrations. From 1999 to 2006, she was a Pacific member of the IFLA Regional Standing Committee for Asia and Oceania and has been associated with the Fiji Association of Women Graduates in various executive positions since its inception in 2003; she is now a trustee of the Association. Elizabeth has worked in all sections of the library, including its valuable Pacific Collection. Under her direct supervision are Readers Services, Pacific Collection and Regional Libraries sections, which are managed by professional staff who work closely with her in the development of the sections and the design and application of policy. She has undertaken work at the USP campus libraries in the Cook Islands, Kiribati, Nauru, Samoa, Tonga, Tuvalu and Vanuatu.

Dr Ivan Ganchev received his engineering and doctoral degrees from the Saint-Petersburg State University of Telecommunications in 1989 and 1994. He is an Associate Professor at the University of Plovdiv and is currently a Lecturer and a Deputy Director of the Telecommunications Research Centre, University of Limerick, Ireland. His current and previous activities include: founding partner and member, ANWIRE (Academic Network for Wireless Internet Research in Europe), the EU FP5 Thematic Network of Excellence IST-38835, 2002–2004; member of two European Co-operation in the field of Science and Technology research Actions (COST 285 and 290). Previous posts held by him include Senior Lecturer in the University of Plovdiv, part-time Senior Lecturer in the University of Shumen and Telecom Expert in Bulgarian Telecom. His research interests include wireless networks, mobile computing and new m-learning paradigms. Ivan has served on the Technical Program Committees (TPC) of a number of international conferences and workshops. He was Track Co-chair of the 65th IEEE VTC2007 Spring conference and TPC member of the IEEE Globecom 2006 conference.

Bob Gann is Head of Strategy and Engagement for NHS Choices at the UK Department of Health. He joined the Department from NHS Direct, where he was Director of New Media, responsible for the NHS Direct website and digital TV services. During his career Bob has worked in healthcare libraries, in NHS public affairs and as chief executive of a not-for-profit agency providing health information services. He has also been a writer and editor for the British Medical Association, and had a regular column in *British Medical Journal*: 'What your patients are reading'. Bob has served on a number of working parties and task forces, and was one of the 25 NHS leaders who signed the NHS Plan. He is Visiting Professor in Health Informatics at Plymouth University, and a Fellow of the Chartered Institute of Library and Information Professionals (CILIP).

Dr Vahideh Zarae Gavgani is a researcher, librarian and Consultant Professor at Tabriz University of Medicine (Iran). She graduated from Osmania University (India). Her PhD research was in Information Therapy (in the field of Medical Library and Information Science). She has published around two dozen research papers in national and international journals and participated in many national and international conferences. She has also published three books, translated a book into Persian and conducted research projects. Her topics of interest are Information Therapy, Health Consumer Information, Evidence Based Medicine, Evidence Based Library Practice, Web 2.0, e-Health, and Medical Informatics.

Hongxing Geng holds an MSc degree in Computer Science from the University of Saskatchewan. His research interests are focused on distributed systems at system level. For his thesis research he developed a mailbox ownership-based mechanism for curbing spam. He currently works for the Library Services of Athabasca University in Alberta, Canada, where he focuses on open access digital repositories design and mobile computing.

Peter Godwin is currently working at the University of Bedfordshire in Luton, UK. Formerly he was Academic Services Manager at London South Bank University, in charge of subject support to all faculties. His interest in information literacy has focused on support to academic staff in universities and the impact of Web 2.0 on information literacy in all information sectors. He has presented widely on Web 2.0 and how it affects the content and delivery of information literacy. In 2008 he co-edited the pioneering book *Information Literacy Meets Library 2.0* for Facet Publishing, which is supplemented by a blog of the same name. He draws on many years' experience in academic library management and has presented at conferences in Europe, Asia, the USA and Canada.

Jim Hahn is the Orientation Services Librarian at the University of Illinois Urbana-Champaign, USA. His librarianship is centred on supporting first-year and transfer students making the transition to university study. He is also involved in library instruction, library tours and reference services. Jim's research investigates the intersection of emerging technology with long-standing LIS practice and how the resulting services will help to support undergraduate students.

Darren James Harkness has a master's degree in Humanities Computing (2008) from the University of Alberta, Edmonton and a BA in English (2002) from the University of British Columbia, Vancouver. He is presently serving at Athabasca University as a web developer and project manager for the Centre for Research. He has previously published a book on the

Apache web server (2004), contributed to a book on intranet design (2003), and has forthcoming publications applying interface studies and post-humanist theory to the production of identity in electronic literatures. His research interests include online culture, electronic literatures, social networking and interface/infrastructure studies. Darren is a member of the Society for Digital Humanities. He has been awarded the Province of Alberta Graduate Scholarship (2005) and University of Alberta Graduate Student Scholarship (2005) for his graduate work.

Dr Anne Hewling is an education consultant specializing in online learning and instructional design. She completed her PhD in educational technology at the Open University and worked there for several years on new technology projects including PROWE, which concentrated on wikis and blogs to support staff development, and on iKnow, which uses web and mobile tools to improve workplace information literacy and management skills. She currently divides her time between education projects in the developing world and teaching both e-learning innovation and international development online.

Maureen Hutchison is Manager of Learning Services for the Centre for Innovative Management in Athabasca University's Faculty of Business. Maureen's current role involves oversight of digital educational publishing activities (project management, editing, instructional design, copyright, course material acquisition/distribution) for graduate and doctoral-level business programmes. Maureen's interests are in instructional design, the use of technology within learning platforms, the creative exploration of learning and business solutions, and management issues. Maureen holds a BA (distinction) from the University of Western Ontario (English Literature) and is in the final stages of completing her MBA from Athabasca University. She has recently co-authored book chapters in diverse areas: future trends in project management (*Project Management Circa 2025*); mobile libraries (*M-Libraries: libraries on the move to provide virtual access*, Facet Publishing, 2008); and online learning (*Theory and Practice of Online Learning*, AU Press, 2nd edn, 2008). Maureen has also contributed to practitioner papers (Sloan Consortium, Magna Publications) involving higher education and business.

Faye Jackson has been working, in various roles, in the library service at Roehampton University, London for 12 years. In 2006 she was appointed to the post of Head of Library User Services. Her work focuses on providing excellent customer service to students and staff of the University. This has more recently led to her involvement in promoting libraries to schoolchildren who may be thinking of applying to university. Faye has also been involved in providing a user-friendly approach to social learning, introducing staff to new and exciting opportunities and challenges in using new technologies to support their teaching and learning through the Green Room at the University Library.

Michelle Jacobs is currently the Emerging Technologies and Web Coordinator at the College Library of the University of California, Los Angeles. She has previously worked as one of the founding librarians at the University of California, Merced. Having worked in a variety of academic libraries and a public library, Michelle focuses her work on the needs of users. Her research explores information-seeking behaviours of undergraduates and developing web content for visual learners.

Kevin Lindstrom is a Science and Engineering librarian at the University of British Columbia, Canada. Kevin has a BSc in Physical Geography (Geomorphology) from the University of Alberta and an MLIS from the University of Western Ontario. He is active in institutional repository developments at UBC.

Graham McCarthy is a Systems Analyst at the Ryerson University Library and Archives in Toronto, Ontario. Over the past two years he has assisted in the expansion of the Library's web presence, developing many innovative and creative applications for the University community. In 2007, Graham received his undergraduate degree in Information Technology from York University, Ontario, and he has recently completed the course work for his master's degree from the Faculty of Information at the University of Toronto. He hopes one day soon to become an academic librarian!

Rory McGreal is a Professor and Associate Vice President, Research at Athabasca University – Canada's Open University. He graduated from McGill University with a joint Honours in History and Russian, and has a BEd degree in language teaching from Dalhousie University, Halifax, NS. His master's degree from Concordia University, Montreal was in Applied Linguistics. His PhD degree in Computer Technology in Education from Nova Southeastern University's School for Computer and Information Science was taken at a distance using the internet. Previously, Rory was the executive director of TeleEducation New Brunswick, a province-wide bilingual (French/English) distributed distance learning network. Before that, he was responsible for the expansion of Contact North/Contact Nord (a distance education network in the remote region of Northern Ontario – serving remote native reserves and small mining towns). He has worked in Arctic Quebec teaching Naskapi Cree students. Rory was the founder of the world's first e-learning website for TeleEducation NB and one of the world's first metadata learning object repositories, the TeleCampus. He has been a leader in the development of the CanCore metadata implementation profile for implementing the IEEE LOM international standard for learning objects. In 2002, he was honoured as recipient of the Wedemeyer Award for Distance Education Practitioner. He has published numerous articles and book chapters on e-learning in Europe, the Americas, Asia and Australia, and has a particular research interest in Open Educational Resources and interoperability, more recently using mobile devices.

Damien Meere received his BSc degree from the University of Limerick in 2005. He is currently pursuing his PhD within the Telecommunications Research Centre at the University of Limerick. His research is focused on the provision of context-sensitive mobile services within a university domain. His interests also include computer networking, development of Java-based intelligent agents and developing advanced mobile learning (m-learning) solutions.

Keren Mills is the Innovations Officer in the Research and Innovations Team at the Open University (UK) Library and Learning Resources Centre. She is primarily responsible for running the Digilab, which is a staff development resource in the Library for Open University staff, giving them the opportunity to get hands-on experience with a range of educational technologies. Keren also undertakes horizon scanning around technologies that could be used to improve library services, such as mobile technologies, social networking software and gaming.

Joseph Murphy is Science Librarian, Coordinator of Instruction and Technology at the Yale University Science Libraries. He earned an MLISc degree from the University of Hawaii in 2006 and received the *Library Journal* Movers and Shakers award in 2009. Joe runs the popular Twitter account libraryfuture.

Gill Needham's current post is Associate Director (Information Management and Innovation) in the Open University Library, where she is responsible for both the acquisition and management of content and the strategic development of new services and initiatives. Since joining the Open University in 1998 she has taken a leading role in developing the library's electronic services to its 200,000 students, has launched and developed an information literacy strategy for the university and has been a major author on several Open University courses. In 2006 she was awarded a National Teaching Fellowship by the Higher Education Academy. Previously she worked for 15 years in the National Health Service, initially as a librarian and then subsequently as an R&D Specialist in Public Health, responsible for promoting evidence-based practice and public involvement in healthcare decision making

Dr Paul Nelson is a public health technologist and Doctor of Medicine. He has a PhD from, and is an honorary senior lecturer at, Imperial College London. In 2005 he founded Phrisk Ltd, a digital public health agency specializing in developing and evaluating digital and internet services for health and well-being. Paul has worked as public health adviser for NHS Choices and has set up two not-for-profit professional and public democratic publishing communities (www.wikiph.org and www.emunity.org) aimed at empowerment through knowledge, connection and contribution.

Dr Máirtín Ó'Droma received his bachelor's and doctoral degrees from the National University of Ireland in 1973 and 1978 respectively. He is a senior academic and Director of the Telecommunications Research Centre at the University of Limerick, Ireland. Current activities include founding partner of TARGET, a European Union Network of Excellence, IST-507893, 2004–08, www.target-net.org. He is a founding member of two European 'COoperation in the field of Science and Technology' Research Actions (COST Actions 285 & 290) focused on simulation and network aspects of wireless communications. He was a founding partner of ANWIRE (Academic Network for Wireless Internet Research in Europe), a European Union Network of Excellence, IST-38835, 2002–04.

Previous posts held by Máirtín include those of lecturer in University College Dublin and in the National University of Ireland, Galway, and Director of Communications Software Ltd and of ODR Patents Ltd.

Dr Mícheál ÓhAodha works as a librarian and a lecturer (part-time) at the University of Limerick, where he teaches on a number of HPSS (History, Politics and Social Studies) courses relating to the history of Irish migration. He has published two dozen books, including: *Irish Travellers: representations and realities* (Liffey Press, 2006) and *American 'Outsider': stories from the Irish traveller diaspora* (Scholars Publishing, 2008). He has a particular interest in the use of technology as a means to circumvent barriers to educational access for groups who have traditionally been unable to access tertiary education.

Victoria Owen is a Project Officer for Learner and Student Support Services at Liverpool John Moores University. She is currently researching mobile technologies in teaching, learning and student support; throughout her research, she has maintained a project blog (http://vickiowensm-learningblog.blogspot.com). Victoria has presented at the CILIP Multimedia, Information and Technology (MmIT) Group Seminar on Mobile Learning in Libraries and at the International M-Libraries Conference 2009 in Canada. Her current work is centred on using mobile communications technologies for student support activities in academic libraries. She has published in CILIP's *Library and Information Gazette* and has publications forthcoming in *SCONUL Focus*. Her main interests are mobile learning, technology-enhanced learning, 'Net Gen' learners and social media.

Alexa Pearce is a Reference Associate in the Social Sciences and Humanities Reference Center at New York University's Bobst Library. She is currently the Acting Librarian for Journalism, Media, Culture and Communication. Alexa is co-chair of the Virtual Reference Services Subcommittee and coordinates the library's text-messaging reference service. She is currently enrolled in a dual-degree graduate programme, in which she will earn an MSLIS from Long Island University and an MA in World History from New York University. Alexa is a member of the American Library Association and the Association of College and Research Libraries. Her research interests focus on the nature and quality of reference communication in online and mobile environments.

Adoració (Dora) Pérez is Library Director of the Open University of Catalonia (UOC). She was Professor in Library Science at UOC from 1999 to 2002, and is the author of online learning materials on the subject 'virtual libraries supporting documentation studies at UOC', as well as of many publications on the topic. She is pursuing her PhD degree on the topic of digital libraries and has already achieved the first step for this, the DEA (Diploma on Advanced Studies). She has degrees in Hispanic Philology and in Library and Documentation Science from the University of Barcelona. She is a member of the executive committee of the Spanish University and Scientific Libraries National network and coordinator of the Research Support Line.

Kate Robinson has worked in many sectors of librarianship – public, corporate (advertising) and even a private member's library – but has now settled in the academic sector and has worked as Head of Academic Services at the University of Bath Library since 1998. Her recent professional interests are around mobile technology, engagement with international students and information literacy from school age onwards. She is a member of CILIP's Chartership Board, Chair of Governors for a First School in Somerset and a magistrate.

Fred Rowland is a Reference and Instruction Librarian for Classics, Economics, Philosophy, and Religion at Temple University in Philadelphia, responsible for research support, instruction and collection development. He is a book review co-editor for *portal: Libraries and the Academy*. Before becoming a librarian he spent many years as a bookseller. He holds an MLS from Drexel University in Philadelphia and an MBA from Temple University.

Adam Shambaugh is a Reference and Instruction Librarian for Business, Finance, and Advertising at Temple University in Philadelphia. Adam provides research, collection development and instruction services to students and faculty members in Temple's Fox School of Business. He holds an MLIS degree and an MA in Linguistics, both from the University of South Carolina.

Hassan Sheikh is currently Head of the Systems Development team at the Open University (UK) Library and manages both systems developers and IT support staff. His role is as technical lead on several internal and external projects, including the continuous development of the OU Library websites, collaboration with several international institutions on the development of mobile library services, shaping up the OU Library Systems Strategy, and advising the OU Library senior management and staff on the latest technology trends in the world of libraries. Hassan has several years of experience in programming and usability evaluation, and worked at the Institute of Educational Technology for four and a half years before joining the OU Library in October 2006. His research interests include evaluating emerging (in particular library-related) technologies, development of mobile library services, programming, usability and user interface design, enabling interoperability between different systems and tracking website user behaviour.

Nafiz Zaman Shuva is Lecturer at the Department of Information Science and Library Management, University of Dhaka, Bangladesh. He has BA(Hons) and master's degrees in the field of Information Science and Library Management from the University of Dhaka, and is currently the Erasmus Mundus Scholar for the 2009–11 academic sessions. Nafiz has more than 10 research publications to his credit, and is the founder president of the Bangladesh Association of Young Researchers (BAYR). His research interests are in digital library system development, mobile applications in libraries and education institutions, e-government and its related aspects.

Brena Smith is currently the Coordinator for Reference, Outreach, and Instruction at California Institute for the Arts (CalArts). Through her teaching and outreach efforts, Brena's work serves the students, staff and faculty of CalArts. She teaches many types of instructional classes, including a credit-bearing course in research methods through the School of Critical Studies. Prior to CalArts, she served as the Information Literacy Operations Librarian at UCLA's College Library. Brena's research interests include innovations in delivering instruction and reference services, and looking at ways scholars from different disciplines utilize the library.

Dr Stanimir Stojanov received his informatics and doctoral degrees from the Humboldt University in 1978 and 1986 respectively. He is an Associate Professor of the University of Plovdiv, Bulgaria and currently

Head of the Computer Systems Department and of the eCommerce Laboratory. His research interests include: service-oriented architectures, agents and multi-agent systems, e-commerce and e-learning applications. Stanimir is a member of a number of international conferences and workshops: PISTA'04, PISTA'05, PISTA'06, SOIC 2005, EISTA'05, EISTA'06 (USA), MENSURA2006 (Spain), and 1st and 2nd Balkan Informatics Conferences.

Tony Tin is the Head of Digital Initiative and Electronic Resources at Athabasca University Library, where he oversees the library's digital preservation, repository development, electronic publishing, mobile library, metadata harvesting service, and other digitization initiatives. Tony holds BA and MA degrees in History from McGill University and a BEd and MLS degrees from the University of Alberta. He has organized a number of mobile learning projects to promote and make the Library's digital collections and learning content more widely accessible and to support distance learners using mobile phones to access library resources and materials. His Mobile ESL project has received an honourable mention for Excellence and Innovation in Use of Learning Technology from the Canadian Network for Innovation in Education (2008). In 2002, he was honoured as recipient of the AU Sue and Derrick Rowlandson Memorial Award for service excellence. He has published articles and book chapters and presented at conferences on topics such as library technology, digital libraries, and mobile learning. His research interests include mobile libraries, digital information management, and digital preservation.

Pep Torn is the Technical Director of the Open University of Catalonia Library. He worked as Internet Librarian at the Technical University of Catalonia from 1999 to 2008. For the last three years he has been the person in charge of the 'Factoria', the media lab of the Technical University of Catalonia. Pep has an LIS degree from the University of Barcelona and a master's degree in LIS from the Open University of Catalonia.

Rachel Vacek is the Web Services Coordinator at the University of Houston Libraries. Her duties include the development and maintenance of the libraries' website and various online services, as well as playing with emerging and open source technologies. She has presented at many local and national conferences, and was a 2007 ALA Emerging Leader.

Kara Whatley is currently Life Sciences Librarian and Head of the Coles Science Center at New York University's Bobst Library. She was previously the Information Services Librarian for Biology and Agriculture at Texas Tech University. She holds a master's degree in Library and Information Studies from the University of Oklahoma and a master's degree in Biological Sciences from Texas Tech University. She is a member of both the Special Libraries Association and the American Library Association, where she is a former secretary of the New Member Roundtable and the current publicity officer for the Association of College and Research Libraries' Science and Technology Section. Kara's professional and research interests include mentoring opportunities for new library professionals and user behaviour in virtual reference environments.

Nicky Whitsed is Director of Library Services at the Open University (UK) where she has led the development of Library Services to the Open University's 220,000 students. Under her leadership the Library Service has developed a new vision for supporting distance learners in the age of digital scholarship and social media including the delivering of services to mobile devices. Her current remit includes University Archives and University Records Management as well as support to Learning and Teaching, Research and the University's e-business. Before joining the Open University, Nicky was Information Manager at SmithKlineBeacham, where she was involved in a major culture change programme, and Librarian of Charing Cross and Westminster Medical School (University of London), where she pioneered the introduction of end-user databases for clinicians and students. Nicky has an MSc in Information Systems and Technology and is a Fellow of the Chartered Institute of Library and Information Professionals (CILIP). She is currently on the editorial board of the *New Review of Information Networking* and the Advisory Panel of the Information Studies Department, University of Sheffield.

Sally Wilson is the Web Services Librarian at Ryerson University Library, Canada. She is currently interested in enhancing users' experiences of the library by integrating its services and resources into students' and researchers' preferred environments. She has an MLS degree from the University of Toronto and an undergraduate degree from the University of Guelph.

Tracey Woodburn has a BA from the University of Northern British Columbia, and is a Researcher with Athabasca University. She has worked as part of the Mobile Learning Team on various projects including: Mobile ESL, Workplace English, Mobile French, and a Pronunciation Practice learning module for mobile devices.

Yang Guangbing is a systems analyst and programmer at the School of Computing and Information Systems, Athabasca University, Canada. With over a decade of industry experience, he has been involved in the development of large-scale software in the finance, education and wireless industries. He received his master's degree in Computer Science from Athabasca University, and is continuing his PhD study at the University of Joensuu's Department of Computer Science and Statistics. Currently, he is working as a programmer in the NSERC/iCORE/Xerox/Markin Industrial Research project. Before joining Athabasca University, he worked for a Canadian software firm as a senior Java developer to design and deploy company's main product 'Wealth Management and Front Office Platform'. He also worked for Siemens China as a software discipline manager dedicated to managing and designing its automotive navigation systems, and for Sun Microsystems as an internship software programmer. He has been a Sun Certified Java Developer (SCJD) and Programmer (SCJP) since 2000. He has also worked as a tutor at Athabasca University and a part-time instructor at Humber College, teaching Java programming and web programming courses.

Foreword

Having been privileged to give a keynote address at the 2009 International M-Libraries Conference, I am pleased to pen a brief foreword to these Proceedings.

M-libraries represent a step change in the evolution of educational technology. Sometimes we are so infatuated with – or bemused by – the diversity of devices now available that we assume flexibility and variety to be the principal virtues of technology. Those are indeed important qualities, but the true essence of technology is to allow us to achieve quality at scale, inexpensively.

As educators, we face a threefold challenge. First, we want to maximize access to education, training and learning opportunities. Second, we want to make the quality of learning as high as possible. Third, we want to do this at the lowest possible cost. Think of these three aims as vectors making up a triangle: Access, Quality and Cost. This highlights the challenge, because with traditional teaching methods this is a cast-iron triangle – it is inflexible.

With classroom teaching, whenever you focus on any one of these three aims (wider access, better quality or lower cost) you will slip backwards on the others. Increase access by making classes bigger, and people will complain about quality. Introduce more learning materials, and costs will go up. Cut the costs of the system, and people will accuse you of both limiting access and damaging quality. The constraints of this triangle have handicapped education throughout history. They explain why so many people still believe that you cannot have quality education unless you restrict access to it.

Technology breaks open this triangle by allowing you to increase access, improve quality and cut costs all at the same time.

This has been demonstrated most convincingly by the world's open universities. The UK Open University, for example, has over 200,000 students, ranks fifth out of 100 British universities for the quality of its teaching, and has lower costs per student – and per graduate – than other universities. Not coincidentally, it and other open universities have been leaders in developing m-libraries.

This educational revolution derives from the fundamental technological principles of economies of scale, specialization, division of labour and machines. M-libraries are a splendid new element in this revolution. They enhance access, they lower costs and they transform the quality of the learning environment for students by freeing them from the constraints of space. This advantage applies as much to students on campus as to those studying at a distance. All contemporary students lead varied, mobile and complex lives. M-libraries allow them to research their academic work properly, wherever they are.

I am particularly pleased to salute the transformation that m-libraries have effected in attitudes to distance education. Not very long ago access to library resources was the Achilles heel of distance learning. Institutions either had to send students all the books and articles they might need – impossible for higher level courses – or arrange for access to other libraries, which were unlikely to have all the resources necessary anyway.

M-libraries have eliminated this problem. Librarians are taking full advantage of a wonderful machine – the internet – but just as importantly, they are applying the other features of technology: economy of scale, division of labour and specialization. I much admire the ways that librarians are using the internet to specialize in the provision of documentation and information to students and faculty, and integrating their solutions into mass systems. It is fascinating to observe the creative ways in which universities are personalizing library support to students in ways that would have previously been impossible.

You are helped by other developments in both technology and values, of which I shall mention two. First, technology and values come together in the Open Educational Resources (OER) movement. This was given a major boost a decade ago, when the Massachusetts Institute of Technology made the lecture notes of its faculty freely available on the web. The ideal of a global educational commons continues to gather momentum. The Commonwealth of Learning and UNESCO are working together to ensure this truly does become a global phenomenon with both North and South, contributing to the pool of OER and drawing from it.

Second, reflection on m-libraries leads naturally to consideration of m-learning. Here there is good news and bad news. The good news is that mobile telephony will be the world's first ubiquitous communications platform. Its major growth is now in the global South, where the mobile is not just a phone but a global address, a transaction device and an identity marker for vast numbers of poor people. It has been spontaneously adopted by billions of people and embedded deep in social consciousness. African peasants paint their mobile phone numbers over their front doors. Indian slum dwellers buy SIM cards to use on friends' handsets. Chinese students spend three months' allowance to buy a handset they can surf the web with.

The bad news is that, although mobile phones are ubiquitous in the developing world, the high cost of usage requires that educators must use them sparingly until rates come down. According to the ICT Development Index 2009, the percentage of Gross National Income (GNI) per capita for 25 outgoing calls in predetermined ratios and 30 SMS messages per month varies from 0.1% for top-ranked Singapore, Denmark and Hong Kong to 60.1% for bottom-ranked Togo. In sub-Saharan Africa, mobile costs are more than 20% of the average annual GNI per capita. In contrast, in the developed world, costs are less than 2% of this figure.

Educators find themselves in a dilemma. On the one hand, mobile phones are a highly valued personal technology. A survey found a third of UK owners would not give up their mobile for a million pounds, young adults see it as a 'critical social lifeline' and most 16- to 24-year-olds would rather give up alcohol, chocolate, sex, tea or coffee than live without it for a month! It is clearly attractive for educational institutions to use such a valued personal device to communicate with students 24/7 and to embed learning in their lives.

So far the sure-fire winners in m-learning in developing countries are SMS for administrative purposes. Reminding students of deadlines, giving words of encouragement and providing bite-sized learning snippets have a beneficial impact on motivation and make it more likely that students will complete and pass the course. However, more extensive applications to learning are expanding, too. The Commonwealth of Learning has teamed up with the University of British Columbia to use mobile phones at scale in rural development for illiterate or semi-literate people.

This Learning through Interactive Voice Educational System, LIVES, automates voicemail-based learning materials to reach thousands of women farmers in their local language or dialect. By linking a learning management system to the mobile phone network, LIVES has strong feedback and performance-tracking features, can accumulate a large database of audio learning materials in various languages and can reach more than 1,000 people at a time.

These are just two examples of exciting new applications of technology. Like the others I mentioned, they allow us to take quality learning to scale at low cost. As the power and sophistication of mobile devices increases and the cost of using them goes down, they will take their place as an important part of the toolkit of distance learning, and of learning generally.

The important work presented in these Proceedings on developing the technology and pedagogy of m-libraries is an important foundation for that future. I congratulate the authors and commend this book to a wide readership.

Sir John Daniel
Commonwealth of Learning

Introduction: a virtual library in everyone's pocket is a step closer to education for all

Mohamed Ally

This book is about bringing the library to the learner using the mobile device that the learner already uses for other activities. The current estimate is that there are over four billion mobile phones in the world, 75% of them in developing countries (*The Economist*, 2009). The biggest increase in the acquisition of mobile phones over the past ten years has been in developing countries. As the use of mobile technology grows globally, libraries are digitizing information for access by anyone, from anywhere and at any time. Hence, the use of mobile devices to access library information will allow everyone to have a virtual library in their pocket.

The role of mobile technology in society

Different sectors of society are using mobile technologies to reach customers and members of society. Mobile banking is allowing people to conduct their banking from anywhere and anywhere using mobile technology. This allows for real-time transactions that benefit both the banks and their customers. In healthcare there is mobile health, where citizens can access health information from anywhere to help prevent illness and to find healthcare advice. Mobile agriculture gives farmers the ability to access information on growing healthy crops, determine market demands for crops and decide on the market price for their produce. This information allows farmers to make informed decisions at the right time to maximize their profits. Mobile shopping puts shopping in the consumer's pocket, to shop from anywhere and at any time. Consumers are able to compare the prices and quality of products using mobile devices before deciding which one to buy and from whom to buy it. Mobile government gives citizens access to government information using their mobile device; hence, government in the citizen's pocket. Mobile fishing enables workers out in the sea and rivers to access up-to-date information on types of fish, the price of fish and market demands for the different types of fish. This provides fishery workers with current information on which to base decisions. Mobile learning, puts the school into everyone's pocket, to learn at their own convenience. Now we have the mobile library, where learners will have a library in their pockets to access information at any time and from anywhere.

People in developing countries have limited transportation options, limited computer resources and limited or no hard-wired telecommunication capabilities. Also, the transportation infrastructure, such as roads, buses and trains is not present or is difficult to access. Because of these limitations people in developing countries are using wireless mobile technology to reach the outside world and to access information. As a result, libraries must make their libraries mobile friendly, such that citizens from around the world can have a library in their pockets. Accessing information and learning materials from libraries will only be a 'pocket away'.

A library in everyone's pocket

This book presents procedures, technology, case studies, best practices and what is happening around the world in mobile libraries to make 'a library in everyone's pocket' a reality. The first part, on mobile development around the world provides information on how mobile technologies are used in libraries around the world to bring the library to everyone's pocket. The chapters in this section provide details on the use of mobile technology in developed and developing countries and in distance education and traditional face-to-face educational organizations. It is important to know what is happening in different parts of the world, since the library of the future will be a global networked library where anyone from anywhere can access information and learning materials from this global library. The global networked library will prevent duplication between libraries and allow for the sharing of resources between libraries to make information access efficient for learners.

The chapters in the second part of the book provide information on the use of mobile technologies in libraries. This section provides details on hardware and software for mobile libraries and on what librarians have learned from using these technologies. It is important for the software systems in mobile libraries to format the information so that learners can access the information on their mobile devices. The system should be able to detect the mobile technology the student is using and format the information for the specific device. This will allow learners from around the world to use their existing devices, rather than having to purchase new devices to access library materials. Future mobile devices must be multipurpose devices so that citizens can use the same device to access library information for learning, information for banking, shopping and socializing, and government and other relevant information. Manufacturers of mobile devices must build multipurpose devices to benefit citizens around the world.

The third part of the book describes specific applications for mobile libraries. As libraries start digitizing information for virtual access, the role of libraries will become very important not only for providing information to learners but also for providing access to learning materials. The learning materials of the future will be in the form of learning objects, which will be stored in electronic repositories for easy access from anywhere and at any time. The learning-object repository of learning materials will be managed by the library of the future. This section describes mobile library applications in mobile learning, science, engineering and service to library users. The chapter on 'Mobile access for workplace and language training' is an example of how the library can be involved in mobile learning; hence, placing learning in everyone's pocket.

Part four provides information on the role of mobile libraries in learning and student support. Because libraries are very efficient in managing information and learning resources, they are increasingly playing a major role in distance education, where learners can access course materials from anywhere and at any time. Hence, academics will have to work closely with libraries to place their electronic course materials in electronic repositories.

The fifth part presents research findings on mobile libraries. The use of mobile technology in libraries is a new field where there is limited research. More research is needed on the interface for mobile access, how to design for different cultures and learning styles, and what are the expectations of the new generation of learners with respect to access to information using mobile devices. Also, design of information and learning materials for mobile technology must make use of the multimedia capabilities of the technology. More research is required on how to design and deliver multimedia materials on mobile technology.

As libraries become digitized, as the use of mobile technology becomes prevalent around the world and as the shift towards open access grows, there will be no hindrance to placing a library in everyone's pocket. Allowing people from around the world to have access to information and learning materials from anywhere and at any time will help to achieve the goal of providing universal education for all.

References

The Economist (2009) *The Power of Mobile Money*, 24 September.

Part 1

**M-libraries: developments around
the world**

1

Where books are few: the role of mobile phones in the developing world

Ken Banks

Introduction – ultra-mobility and digital youth

For every personal computer in a developing country there are roughly four mobile phones. Although many of these are likely to be older or low-end models, today's high-end devices have the equivalent processing power of a personal computer from the mid-1990s. In comparison, personal computers in 2009 have more number-crunching ability than all the computers that took the Apollo rocket to the moon over 30 years earlier in 1969. The boundary between mobile phones, hand-held game consoles, entertainment devices and personal computers is becoming increasingly blurred, with devices such as the Blackberry, Symbian-driven smart phones, GPS-enabled mobile phones, Ultra Mobile PCs and Nokia's N-Gage breaking new ground. Technology continues to advance at a remarkable pace, opening up new opportunities few people would have considered a few years ago.

Mobility – and increasingly 'ultra mobility' – is the buzzword of the day. According to the chief executive of OQO, a manufacturer of ultra-mobile PCs, 'Ultra mobility is the ability to access all of your information, get in touch with anyone you want to, collaborate with anyone, and run any application you want from anywhere on the planet.'[1] Convergence is making this possible, with music players, Wi-Fi connectivity, video cameras, GPS units and live television capable of running on a single device, often a mobile phone. The days of carrying a separate phone, camera and music player are over. Indeed, many people are beginning to question use of the word *phone* at all, preferring to refer to these new gadgets as mobile communication devices, or digital assistants.

'M-learning' is a term regularly used to describe the many possibilities opened up by this convergence, whether they be getting exam results by mobile phone, lecture podcasting via iPod or structured language games on a Nintendo. These are still early days, and while examples of m-learning in action are continually on the rise, the benefits have already begun to be studied and documented. In *Mobile Technologies and Learning: a technology update and m-learning project summary*, Jill Attewell (Attewell, 2005) highlights several. According to her findings, m-learning is helping students to improve their literacy and numeracy skills, and to recognize their existing abilities. It also encourages both independent and collaborative learning experiences, and helps learners to identify areas where they need assistance and support. It can help to combat resistance to the use of ICT, which can help in bridging the gap between mobile phone literacy and ICT literacy, and help to remove some of the formality from the learning experience which engages reluctant learners. And it can help learners to remain more focused for longer periods.

Further studies are painting a picture of today's youth becoming increasingly comfortable and accepting of their new digital lifestyles, powered by always-on technology such as mobile phones, and enriched by portable entertainment devices such as iPods, digital cameras, Sony PSPs and the Nintendo Game Boy. Friendships are made, maintained and lost online, often in virtual worlds and on social networking sites such as MySpace, Facebook and Bebo. Much of what we are seeing today – generally out of the classroom but increasingly in it[2]– is technology driven, but this technology is not universally accessible.

Mobile access in the developing world

The living and learning environment in developing countries can often be quite different. Where mobile technology may prove a complementary extension to teaching methods in the West – improving or enriching the learning experience, for example – in many developing countries it offers the hope of revolutionizing learning altogether, even taking it into areas previously starved of reliable or regular education services. This is particularly true in rural areas, which may be characterized by a lack of fixed telephone lines, poor roads and unreliable electricity, poor postal services, few if any personal computers, few teachers and most likely no internet access.

What many of these communities will have, however, is mobile network coverage and, if not their own phones, at the very least access to one. Learning by distance is nothing new in many developing countries, and the mobile phone has the potential to unlock it yet further, expanding its reach and delivering richer, more appropriate, more engaging and interactive content.

But despite the promise, problems remain.

Imagine two mobile phone users. One lives in the Land of Plenty and owns an iPhone. He or she can access the internet via free wireless connections dotted around the city, download and play games, keep in contact with friends and family via instant messenger (IM), watch streaming video and live TV and use as much data, SMS or voice as s/he likes with a cheap, all-inclusive price plan. The other lives in the Land of Less. He or she uses a shared phone, lives in an area not covered by a data network of any kind, has a sporadic signal, a phone technically incapable of playing games or video, and has to think twice before sending an SMS or making a voice call because of constant concerns about airtime credit, not to mention worries over how the phone will be recharged if the mains electricity goes off and doesn't come back for days.

Closing the digital divide by mobile

During spring 2007, I was invited to the 16th International World Wide Web Conference in Banff, Canada. I was there to take part in two separate tracks, although the topic was the same – how the mobile phone might help to close the digital divide in the developing world. My talk on the first day was more general, discussing the delivery of targeted information: health messages, wildlife alerts or market prices, for example – via text message (SMS) – and the importance of understanding the complex cultural issues that surround technology adoption in places like Africa, a place where I have done most of my work. On the second day I sat on an Expert Panel discussing something a little more specific: access to the internet via mobile devices under the same developing country conditions.

I started my Panel discussion with a short description of what I considered Utopia, the ideal conditions under which we'd all like to be working. It went something like this:

> Everybody, everywhere, wirelessly communicating and accessing a whole range of personally relevant information whenever they like using a wide variety of compatible devices at high speed and low cost.

This, of course, isn't realistic *anywhere*, let alone in many developing countries, at least not yet. But the specific problems of web delivery in these places are not dissimilar to those faced by anyone looking to work with mobile technologies in the developing world. And, as you would expect, the m-learning community is not exempt. Ageing handsets, limited functionality, lack of bandwidth, issues of literacy and cost are just some of the barriers, and there are many. It is these barriers that I propose to discuss a little later in this chapter.

But for now let's imagine, for one moment, that we *are* living in Utopia and almost anything is possible. The sky's the limit! What would that look like? Given a high-end mobile device – mobile phone, personal digital assistant (PDA), pocket PC, even things like iPods – what could we do? More to the point, what would *students* require it to do to make their learning experience more engaging, enjoyable and productive, assuming that these are key objectives? Would their mobile learning experience be largely based on video lectures? Collaboration with other students via online blogs and wikis? Playing games and 'learning by doing'? Schooling in a virtual world with virtual classmates, teachers and desks? Pitting students against one another through online spelling and maths competitions? Mobile-delivered examinations? All of these? More?

Online learning made mobile

Some of these things, of course, are already happening. The University of California in Berkeley recently began posting entire lectures on YouTube[3] and, of course, YouTube content is accessible via mobile devices. A lecturer at Bradford University in the UK early in 2008 went as far as abolishing traditional lectures altogether in favour of podcasts,[4] in his words 'freeing up more time for smaller group teaching'. And children can learn to count, spell or even play guitar using Java-based mobile games, downloadable from the internet or directly onto their phones via a carrier portal.

The closer you are, of course, to the optimum device and network conditions the more things become possible. Three projects highlighted below take advantage of some of these optimum conditions, but use the technology in slightly different ways and aim at subtly different target audiences. The first, wildlive!, sets out to raise awareness of wildlife conservation among the wider general public, whoever and wherever they may be. The second, Freedom HIV/AIDS, was more specific, targeting members of the public particularly at risk from contracting the disease, largely in developing countries. And the third, Dunia Moja, is a lecture and class-based education tool aimed at a controlled group of students taking a particular university course.

wildlive!

As 2002 came to a close, a visionary team at Fauna & Flora International – a Cambridge (UK) based conservation organization – began looking at ways in which emerging mobile technology could be used to promote their international conservation effort. A new breed of handset was coming onto the market, with colour screens, internet access, video capability, cameras and the ability to play games. wildlive![5] was launched in the UK in 2003 and then across Europe in 2004, and adopted a combined web and WAP (Wireless Application Protocol) approach, meaning that it provided conservation content on the internet *and* mobile phones. News, diaries, discussions and other information were added to the website, which was then in turn rendered for mobile devices accessing via the Vodafone network. A 'community of interest' was created, allowing users to contact others with similar ideas and views, and a wide range of conservation-based resources and downloads were made available online. Among this innovative range of content were five mobile games which taught users about gorilla, turtle and tiger conservation while they roamed around a range of environments. Another game was a 500-question quiz based on zoology and biology. The project received considerable attention, was nominated for an award and is still seen as groundbreaking today.

Freedom HIV/AIDS

Originally developed for the Indian market, Freedom HIV/AIDS[6] was launched on World AIDS day 2005 and sought to use mobile phones to take HIV/AIDS education to the masses. A number of games were developed, including 'Penalty Shootout' and 'Mission Messenger'. In the shootout game, the player is given points for saving penalties and receives tips on how to avoid HIV/AIDS transmission. At the same time the game seeks to dispel myths surrounding the disease. In the second game, the player 'flies' across the African continent distributing red ribbons and condoms, spreading messages of HIV/AIDS awareness, prevention, transmission and safety. The games were originally developed for the Indian market but have since been translated into a number of African languages.

Dunia Moja

Dunia Moja,[7] Swahili for 'one earth', seeks to use 'mobile technologies to connect international students and faculty to stimulate learning and debate in environmental sciences'. This innovative project, piloted in 2007, was a collaboration between Stanford University and three African academic institutions – the University of the Western Cape in South Africa, Mweka College of African Wildlife Management in Tanzania and Makerere University in Uganda. The project used high-end PDAs to allow students to download and watch video lectures from academic staff in each of the partner universities, and to contribute to the discussion and debate through mobile blogging to a central website. The course was centred around global environmental issues and their impact on the African continent and the United States, and brought local perspectives and viewpoints to bear on the course topics. Faculty and students from the four participating institutions electronically shared course materials, exchanged information and contributed their own course content. In m-learning in developing-country terms, Dunia Moja is a pioneering first.

As these three interventions show – and there are many more out there – much is possible if you have higher-end devices and a fast, reliable data network at your disposal. In the Land of Plenty the sky really is the limit. In the Land of Less, however, we have fewer choices.

M-learning constraints for the developing world

Furthering the advance of m-learning in developing countries is governed by a combination of five key constraints, four of which are technical. There are other non-technical ones such as literacy, culture and language not covered here. Depending on the target area, none or all of these may apply. I consider these technical issues to be as follows.

Mobile ownership

Although growing at a phenomenal rate, mobile ownership in many developing countries is still relatively low, and nowhere close to the near-100% penetration rates that we see in many mature markets. If educational establishments begin to embrace mobile technology to any significant extent, then issues of ownership and access to handsets by students need to be addressed to ensure that, in the words of a recent American President, 'no child is left behind'. Putting learning tools in the hands of children in developing countries is a key objective of the One Laptop per Child (OLPC) project, run by MIT Media Lab, Cambridge, MA. Many people believe that the mobile phone would be a better tool to work with. The debate continues.

Mobile technology

Where pupils do own or have access to mobile phones, more often than not – and this is particularly the case in many rural areas – these phones will more likely be either older model, or lower-end handsets with limited functionality. In order to develop appropriate teaching tools, the reality of the target market needs to be considered. Ownership and use of PDAs and pocket PCs should most likely be considered non-existent among the wider community.

Network access

Higher-end handsets with data capability are only useful in areas where the mobile network can service them, and where costs of data access are not prohibitive. In many cases neither of these is a safe bet. By way of example, during a recent month-long visit to Uganda working with Grameen, for approximately 90% of the time I was unable to connect to the internet using my phone.

Device limitations and lack of industry standards

Mobile phones may be ubiquitous, highly portable, shareable, immediate and always on, but there are also limitations which present challenges to even the most talented mobile applications developers – small (and generally low-resolution) displays, awkward text input methods, slow data access (if at all) and issues of battery life, among others. On top of all that, the mobile industry has historically suffered from a lack of standards – different manufacturers supporting differing video and audio formats, no standard screen size and resolution, lack of regular support for Java and/or Flash, incompatible browsers (if at all) and a wide array of memory sizes. All of this makes the platform landscape fragmented enough to make developing an m-learning application a real challenge.

Despite these issues, however, there is still much that can be done. Text messaging, or SMS, is universally available to mobile owners the world over, and is relatively cheap, direct, and gets around many issues of literacy. A number of African countries allow students to obtain their exam results by SMS, or to check whether they have successfully enrolled on a course – although these examples are based more in the administrative side of education.

M-learning technology in the developing world

In 2005 the University of Cape Town trialled the use of mobile phones to help administer a number of its courses.[8] Text messages were sent out to students whenever rescheduling and cancelling of classes was necessary, whenever there were computer network problems, and when test results became available. According to a spokesperson at the university, 'At a superficial glance, with its concentration on administrative functions, the project does not seem remarkable, particularly as the developed world moves into sophisticated mLearning. The importance of the project, however, is that it illustrates a set of principles useful for the introduction of this technology into the third-world environment, or into any institution making first steps into mLearning.'

In other African countries SMS is being used to alert parents if their children haven't turned up for school, or by children who find themselves the victims of bullying. During an online discussion towards the end of 2007 about the potential of mobile technology in e-learning, a number of initiatives were discussed, including the texting of homework to students, or the ability for students to text in their homework answers and for SMS to be used as a reading aid. With some children living far away from their nearest school, such initiatives could be revolutionary. And with products such as FrontlineSMS,[9] implementing such projects need not be expensive or technically out of reach. Today it is more about 'blue sky thinking' than about the sky being the limit. But it will not always be this way.

Conclusion

Ironically, technological conditions aside, m-learning is particularly well suited for use in developing countries. M-learning is useful as an alternative to books or computers, something generally in short supply. It is empowering in situations where students are geographically dispersed, again a common scenario, and is particularly helpful in getting students up to speed who may have previously felt excluded, or who find themselves behind and needing to quickly catch up.

Mobile technology has revolutionized many aspects of life in the developing world. The number of mobile connections has almost universally overtaken the number of fixed lines in most developing countries in the blink of an eye. If further evidence were needed, recent research by the London Business School found that mobile penetration has a strong impact on GDP. For many people, their first-ever telephone call will have been on a mobile device. Perhaps, in the not-too-distant future, their first geography lesson will be on a mobile, too.

Notes

1   Ultra-mobile Future Beckons for PCs,
     http://news.bbc.co.uk/1/hi/technology/7178278.stm
2   Doug Belshaw's teaching-related blog,
     http://teaching.mrbelshaw.co.uk/index.php/2006/09/21/20-ideas-getting-
     students-to-use-their-mobile-phones-as-learning-tools/
3   UC Berkeley First to Post Full Lectures to YouTube,
     www.news.com/8301-10784_3-9790452-7.html [accessed 3 October 2007].
4   Podcast Lectures for Uni Students,
     http://news.bbc.co.uk/2/hi/uk_news/england/west_yorkshire/5013194.stm
     [accessed 26 May 2006].
5   wildlive!,
     www.kiwanja.net/wildlive!.htm

6   Freedom HIV/AIDS,
    www.dgroups.org/groups/oneworld/OneWorldSA/index.cfm?op=dsp_
    showmsg&listname=OneWorldSA&msgid=498250&cat_id=513 [accessed 30
    November 2006].
7   Dunia Moja,
    http://duniamoja.stanford.edu.
8   www.usabilitynews.com/news/article2572.asp
    www.oucs.ox.ac.uk/ltg/reports/mlearning.doc
9   FrontlineSMS,
    www.frontlinesms.com

Reference

Attewell, J. (2005) *Mobile Technologies and Learning: a technology update and m-learning project summary*, Learning and Skills Development Agency.

2

Mobile technology in Indian libraries

Parveen Babbar and Seema Chandhok

Introduction

The total number of mobile phone users has risen to 350 million since the inception of mobile use in the country, and India is one of the most attractive markets for mobile telephone operators and wireless equipment vendors. The Indian government is targeting 500 million telephones, both fixed and wireless, by 2010. The market is growing by about ten million new mobile users per month, and with this pace of growth India will probably be close to 450 million subscribers by the end of 2009. Four factors have driven the growth of the mobile subscriber base in India. These include footprint expansion by existing operators, especially in rural areas, launch of operations by newer operators, issuing of 3G licences – which opens up a new world of data services – and cheaper handsets, which will lower entry barriers even further (Jain, 2008).

Libraries and documentation centres in India are keeping this proliferation in view and have also initiated some mobile learning (m-learning) services. E-learning has been successful, and now m-learning represents the next stage in an ongoing continuum of technology mediation. It will require new digital communication skills, new pedagogies and new practices. It will also foster young people's interest in their mobile phones and other hand-held communications/entertainment devices to deliver exciting and unusual learning experiences and related messages.

M-learning, especially its main delivery system, mobile-phone learning, is under observation, and in coming years will be effectively and extensively used or accepted for learning purposes by either educators or the general public in India. The goal of modern digital libraries in India is to support 'nomadic' computing by providing appropriate wireless networking 'hot-spots' and access to information through mobile devices to support flexible learning space and mobile learning.

The proliferation of mobile technology over the past decade has made librarians face issues that were previously the responsibility of software designers and specialists in the field of human–computer interaction (HCI). In this chapter, we will describe the weaknesses inherent in, and some non-viable factors of, mobile phone learning. We will describe how libraries can combine the technologies of mobile communications with any electronically delivered material to impart support services, and will discuss the new PDA technologies, smart phones and wireless connections available in India. This chapter will also present strategies for delivering educational resources to mobile devices through libraries in India.

The 3G path to m-learning

The Telecom Ministry of India is responsible for the roll-out of third-generation (3G) mobile technology called '3G mobile services', which will be available in India by mid-2009. These 3G or next-generation mobile communications systems will enhance mobile services and will offer multimedia, internet access and the ability to download and view videos. The technology will enable network operators to offer users a wide range of more advanced services while achieving greater network capacity through improved spectrum efficiency.

With everyone and everything going mobile, why not learning too? It is the natural progression of technology. This exciting use of technology enhances the learning experience. Hence, m-learning is hardly a surprising development in this super-technology era. It is actually e-learning through hand-held computational devices. These devices can easily be transformed into tools for accessing courses through online content delivery systems, or from offline library management systems (LMSs) stored locally in the handsets.

M-learning is ideal for those who are always on the move, tech-savvy, know what they want and when they want it, and don't want to stop. M-learning is a revolutionary idea to widen access to learning, and which aims at not only reducing social exclusion but accelerating development of the workforce and encouraging healthy competition. It can be accomplished on PDAs, smart phones and other similarly enabled mobile devices. These can support various languages, graphics and animations, and are equipped to manage data on the move.

Today's trends indicate that mobile phones are being transformed into classrooms on the move, providing information on demand via text, infographics and video. In simple terms, m-learning is the art of using mobile technologies to enhance the learning experience. The service utilizes tools such as digital content from traditional textbooks, and online databases, amongst others, through mobile phones to provide a dynamic learning environment.

High-quality mobile learning education and training materials can be tailored to suit individual student needs and made available whenever and wherever they are required. This innovative style and technique enhances dynamic learning, where content is custom-assembled and delivered to students according to personal pace and need. The declining costs of mobile handsets and services as compared to laptops, and the likely elimination of the need to be physically connected to the world wide web, with advent of 3G, will provide an impetus to m-learning in India.

Indian libraries will in the near future be offering amazing services via the mobile web. Imagine where we will be in a year or two, as mobile internet uptake continues to increase and portable devices steadily improve. Some of the advanced applications to be included in Indian libraries are:

- Students entering the library stacks will be able to scan a barcode to access a navigation guide to the layout of each floor.
- Library patrons will be able to check out their own books and media items. This might become possible with mass adoption of 2D barcode readers, similar to those available in conjunction with the virtual wallet capabilities that are currently being adopted in Japan.
- Via mobile access mobile phone owners will click on a library icon offering them shortcuts to desired library content such as e-books and audio books, without ever having to open a web browser.
- Clicking on a mobile phone icon will initiate a video conference with a research librarian. With powerful services such as Skype Mobile, this has become a reality.

Libraries and the concept of the mobile web

The mobile web is nothing more than the world wide web made accessible through a mobile device ranging from a mobile phone to an iPod Touch. It constitutes the entirety of the internet and is not limited to websites that have been specifically designed for mobile viewing. Handsets and mobile phones that are web enabled can search and browse the internet from anywhere.

Mobile websites are made especially for the small screen. They appear as scaled-back versions of their desktop counterparts, mostly with a numbered menu system for quick access to content. Web pages that do not have mobile versions appear as if they have been squeezed onto the tiny screen, with overlapping menus and links. A website can also be 'transcoded', or formatting can be applied to make it more readily viewable on a phone (Kroski, 2009).

There are seven major reasons why libraries should go mobile:

1 There are three times as many mobile phones in the world as personal computers.
2 Mobile makes your content ubiquitous.
3 Mobile diversifies your audience.
4 Mobile enables you to offer new service types, i.e. location-based.
5 Mobile enables you to connect to patrons via a new medium.
6 Mobile is the way of the future.
7 It's easier than you think.

Transcoded websites

Transcoding is a technology that takes a regular website and reformats it for display on a tiny mobile screen. When using a mobile device, many search engines, such as Google, will show transcoded versions of web pages as results, along with any mobile editions of the site. But developers, as well as users, can transcode websites directly, a free transcoding application such as Skweezer or Mowser which compresses the HTML content of a website to produce a single column. These transcoded web pages are viewable on a wide range of mobile phones; however, the automatic nature of the web page transformation often results in excessive scrolling and less-than-perfect displays.

Mobile-only websites

Designing a mobile-specific website provide not only more freedom in the design, content and structure of a portable web page, but additional choices about the type of technology and format to develop it in. XHTML-MP and Wireless Markup Language (WML) programming languages allow developers to create robust mobile websites. There are applications such as Zinadoo and dotMobi's Site Builder which provide FrontPage-like development interfaces for creating mobile websites from scratch. Similarly, many free applications are available to help organizations create their own mobile websites, such as Winksite and Mofuse to create a mobile version of websites from RSS feeds. These applications also provide tools to create Quick Response barcodes and widgets that can be added to desktop websites offering to send the mobile URL to visitors who enter their phone numbers. Furthermore they can embed code for adding the site to blogs or other websites, iPhone-only websites, and links to share websites with social networks and communities.

Benefits of the mobile web

The mobile web is the internet for the small screen, and thus provides many of the same benefits as its desktop counterpart, such as:

Constant connectivity: Web-enabled mobile devices provide owners with around-the-clock access to the internet, regardless of location.

Location-awareness: Many of today's smart phones and pocket PCs have global positioning system (GPS) capabilities which make them aware of where they are at all times.

Limitless access: The mobile web encompasses not only those sites that have been specially designed for mobile browsing, but also the world wide web. Web-enabled phone users have access to all of the same online resources that they would find via their desktop computer.

Interactive capabilities: The mobile web offers users the participatory experience of the read/write web in the palm of their hand. Users can create content, share and rate media, make comments, write blog posts, tag resources, and form connections on social networks (Kroski, 2009).

Limitations of the mobile web

The mobile web also has some limitations:

Slow connectivity: Generally slow connectivity is a major issue. To overcome this, mobile web is offering content as downloadable modules that can be transferred to the mobile device using Bluetooth or a USB data cable.

Data costs: Data costs are high at present and connectivity is slow. GPRS (General Package Radio Services) is the norm, with EDGE (Enhanced Data rates for GSM Evolution) coverage being spotty, and 3G is still in limbo.

Multiple standards: Multiple standards come in different mobiles, with different screen sizes and operating systems.

Repurposing existing e-learning materials for mobile platforms: It is a complex activity to restructure existing e-learning materials for mobile platforms.

Display of large digital content: Sharing of high-resolution images and uploading of PowerPoint presentations can prove difficult, taking into consideration the small size of the devices.

Devices for the disabled: Content provision for the reading impaired will be another challenge.

The m-learning scenario in India

People in India have an insatiable thirst for information and knowledge. Moreover, mobile services in India are quite affordable; hence, even an ordinary person can own and use a mobile phone. Added to this is the fact that India happens to have one of the largest populations in the 18 to 28 years age group. M-learning in India is at present still in its infancy. However, the future promises an exponential market. The proliferation of mobile phones, PDAs and other mobile devices means that the platform has a lot of potential in India, with over two million users being added every week and a total of around 300 million mobile users in 2009, and excellent connectivity across regions. Although the greater part of this user base is not using advanced devices required for effective multimedia-based m-learning, the figures are too high to be ignored, considering the interest in and growing number of 3G devices (Prakash, 2008).

Major mobile manufacturers such as Nokia, Sony Ericsson and Motorola in India have linked up with service providers like Airtel, Vodafone and others to provide mobile content, which also includes learning content. Companies that specialize in content aggregation provide the actual content, while mobile value added service (VAS) providers develop the mobile technology and delivery. Most of these companies have already launched their services and even the standard Graduate Record Examination (GRE) examination papers are now available on the m-learning platform.

In view of this, Hewlett Packard has awarded a 'Technology for Teaching' grant to Jadavpur University in Kolkata to transform teaching on the campus. The university has received HP tablet PCs, external storage and optical drives, wireless networking cards and printers, as well as funding for staff to work on the project. This will help the university to establish an m-learning centre where students taking the MTech course in Distributed and Mobile Computing can access content using hand-held devices. The university already has a digital library, and a content management and development system using an m-learning authoring tool. Students will be able to connect to a server-based open source wireless laboratory, built on existing laptop computers and wireless technology. Another university that has been selected by HP for this global award in India is Anna University in Chennai. Similarly, an IT training institute in India, Aptech Learning Services, has also developed an m-learning platform to cater to the educational needs of corporations and institutions.

Mobile technology in the library

'Libraries in Hand' is the latest slogan for Indian libraries. Indian library professionals are on their way to determining how these devices can help with information access and ensure that they can communicate with patrons and provide web content in the most appropriate and effective ways. Many efforts have been made to popularize their services and to increase the market and demand for mobile access to personalized facts and information anytime, anywhere on one's own hand-held device. Indian libraries are all set to use many of the technologies made available by the mobile industry, such as PDAs, Blackberrys, iPods, mobile phones, ultra mobile PCs (UM PCs); to mobilize library content in portable formats suitable for small screens; and to deliver library alert or information services in the form of contents/information compatible with devices' range of search features. Libraries are mastering the mobile web so as to bring patrons a new set of services. They are exploiting the technology that their patrons are currently using, such as mobile phones and iPods, to deliver robust new services without the users having to leave their comfort zones. These portable offerings are serving to integrate library services into patrons' daily lives.

Mobiles in the stacks

Library professionals need to become proficient in using these devices so as to enable users to access them anywhere, at any time. One of the major examples is a newly released Sirs product called Sirsi PocketCirc. This is a piece of software which runs on a PDA and allows library staff to perform circulation tasks in any part of the library that has wireless connectivity. This circulation solution helps libraries to extend services, increase operation efficiency and save staff time. It merges the simplicity of a handheld device with online and offline circulation functionality in a Windows environment. The pinpointing of book locations in the library uses the mobility of the device to full advantage and adds another degree of self-sufficiency to the transaction. Libraries may want to consider providing access to circulation records, book due dates, overdue notices and interlibrary loan (ILL) requests via mobile phones and hand-helds so as to better serve their mobile patrons. It is a boon to the staff, as it frees them to serve users and perform both online and offline circulation operations without having to be at a desktop workstation (Khare, 2009).

Mobile literature

Another venture is being undertaken by VIT University, Vellore. MobileVeda, an organization developed at the Technology Business Incubator, VIT University, has launched current literature (a collection of short stories and poems) in the form of mobile books. A short-story collection by the writer Bharathibalan and the poems of the poet Kalyanji were launched as mobile books. MobileVeda will also be launching a microchip (mobile memory card) containing a thousand books that can be read on a mobile phone, which is to be unveiled by Thiru G. Viswanathan, Chancellor of VIT University (Viswanathan, 2009).

Future mobile services

Some of the services that Indian libraries need to implement in the near future are:

Mobile library websites and MOPACs (mobile OPACs): Libraries are creating mobile versions of their websites for their patrons to access on the go. These libraries will be offering information about library services and collections, access to library catalogue search, portable exhibit information, subject guides, e-journals and library hours, all formatted for the small screen.

Mobile collections: Libraries can also offer their patrons digital media collections that they can take to go, enabling them to benefit from library services remotely. These can include audio book collections, e-books, video and music files. Patrons can transfer a wide range of media items to their cell phones.

Mobile library instruction: Library users who don't have the time or inclination to attend an on-site workshop can still get the most out of library resources by accessing classes and tutorials on their mobile devices. Libraries can distribute their knowledge of and expertise in library systems and materials via MP3 and video files that patrons can take with them. A series of short audio files can be created describing the library, how to get reference assistance, and library workshops.

Mobile databases: It's not only libraries that have seen the writing on the wall with regard to the mobile web, but academic software and database providers have started taking portability to heart. Many scholarly database management applications are providing search interfaces for mobile web users.

Mobile audio tours: Libraries can make guided tours more convenient for patrons with busy schedules by making self-service audio tours available for hand-held devices. Rather than asking patrons to schedule an appointment in advance, or learn to utilize a new technology, these new audio tours can make the most of patrons' MP3 players and mobile phones to impart information.

Library SMS notifications: Text message alerts provide busy mobile owners with quick news announcements, reminders about important events, or requested information. Libraries can offer these speedy advisories as an added service to patrons.

SMS reference: Reference services in libraries today are becoming increasingly virtual, as more and more researchers are working remotely. Technologies such as instant messaging, e-mail and SMS text messaging are making it easy for libraries to maintain their relevance as information hubs by offering convenient services to busy users. Ask-a-Librarian services can be offered to mobile patrons, enabling them to submit their research questions remotely, by text.

Mobile library circulation: Not all new mobile tools involve direct patron interaction: some can be used behind the scenes to offer improved library services. SirsiDynix has developed a hand-held circulation tool called PocketCirc, which enables librarians to access the Unicorn Library Management System on a PDA device. This wireless application enables staff to assist patrons in the stacks, check out materials while off site, such as at community or campus events, and update inventory items while walking around the library.

Future potential of mobile applications in libraries

More and more changes are expected within four to five years in the field of mobile technology and its application to libraries. The technology is now available to use phones to read barcodes or RFIDs (radio frequency identifications) in the library, and OPACs are developing GIS (geographical information system) sensitivity and the ability to communicate with users through their mobiles for reservations, fines, late notices, alerts, etc.

In the near future we expect to have large components of asynchronous voice messaging, including threaded discussions using voice-mail technologies that will assist library staff in providing ready reference service, leaving behind texting and SMS. Timed voice-mail as well as mobile voice-blogging will greatly enhance the usage of mobiles in future. Mobile Web 2.0 and 3.0 applications for social networking for the library community are available, thus enabling discussions, blogs, wikis and other features beneficial for all library developments.

Privacy and copyright are major matters of concern, due to the availability of web content 24/7 and the possibilities both of its corruption and loss on the computer and of mobile searches by individuals without any authentication or identification; this is going to be the fastest growing application in the next five years. Librarians have to demonstrate a full understanding of the capabilities and potentials of mobile technology and its use in libraries in the near future, by providing quality-based services matched to the needs of the user.

Libraries in India need to consider the provision of content and services for mobile users on two levels: internally, within the library, and at an institutional level. Some issues that the library may wish to examine in house are the library's role in:

- licensing information products for mobile devices
- hosting or pointing to institutional content intended for mobile devices, e.g. podcasts
- preserving new content types and formats
- providing instruction on the devices themselves, not just access to content
- providing space for new equipment and work styles.

Similarly, libraries should take a campus leadership role and consider establishing a task force or study group that includes individuals representing various sectors of the institution to examine issues related to mobile users; or if such an institution-wide group already exists, libraries may want to ensure that it is represented. The group may want to address:

- specific goals and objectives for mobile content/services
- the current state of uptake of mobile devices among patrons
- target audience for anticipated content/services
- interested parties who should be involved in the detailed planning
- a clear understanding of resources needed and funding streams
- a plan for assessment of the effectiveness of the new content/services.

As with most technology developments, mobile technology is being developed at a rapid rate. Thus, Indian libraries need to make conscious choices about what they want to offer in this arena, and act accordingly (Lippincott, 2008).

Conclusion

India may well be one of the leading countries in the adoption of m-learning in coming years, owing to the numbers of young users or 'Gen Y' involved in multimedia mobile usage. The Indian educational industry is evolving. The shift from 'd-learning' (distance learning) to 'e-learning' and now from 'e-learning' to 'm-learning' will be the next big wave, which will revolutionize education in India. M-learning will bring about a paradigm shift from the traditional methods of education delivery, and integrate ICT as an essential component in everyday learning. Web applications such as Google, Facebook and YouTube have gone mobile, thereby underlining their popular appeal. Following the same trends, it is also possible to develop an m-library presence with relatively little effort. Indian libraries need to be indispensable to their users, and to this end they have to include mobile devices as part of their strategic thinking.

Mobile libraries have to grow, and this requires greater collaboration between academia, industry, corporations and government. In the current scenario, mobile libraries have the potential to proliferate and we will witness a situation in which the mobile will definitely be used as a tool to spread learning across the country.

References

Jain, R. (2008) Emergic: Rajesh Jain's Blog, *India Mobility Trends*,
    http://emergic.org/2008/12/16/2009-india-mobility-trends/

Khare, N. (2009) Libraries on Move: library mobile applications. In: *7th International Convention on Automation of Libraries in Education and Research – CALIBER 2009, held on 25–27 February 2009, organized by INFLIBNET Centre, at Pondicherry University, Puducherry.*

Kroski, E. (2009) *On the Move with the Mobile Web: libraries and mobile technologies*, www.ellyssakroski.com

Lippincott, J. K. (2008) *Mobile Technologies, Mobile Users: implications for academic libraries*, Association of Research libraries, Bimonthly report no. 261, December, www.arl.org/bm~doc/arl-br-261-mobile.pdf

Prakash, N. (2008) m-Learning supplements e-Learning, *Express Computer Online*, (17 November),
    www.expresscomputeronline.com/20081117/market03.shtml

Viswanathan, T. G. (2009) *Launch of Mobile Books and Introduction of Library on a Memory Chip*, VIT University Open PR: Worldwide Public Relations, www.openpr.com/pdf/65265/Launch-of-Mobile-Books-and-introduction-of-Library-on-a-memory-chip-by-Thiru-G-Viswanathan-Chancellor-VIT-University.pdf

3

Mobile technologies and their possibilities for the library, University of the South Pacific (USP)

Elizabeth C. Reade Fong

Introduction

This chapter discusses the way in which mobile technologies may be used to enhance services to users in the context of the Library at the University of the South Pacific (USP), based on a survey of library users at three campus libraries of Alafua (Samoa), Emalus (Vanuatu) and Laucala (Main Campus, Fiji). Background information is provided on the unique setting of the USP and its library network, and is followed by an analysis of the findings of the survey and its implications and challenges for the USP Library.

Setting the scene: the Pacific Islands and the University of the South Pacific

The University of the South Pacific is located in a region spread across 33 million square kilometres of ocean, an area more than three times the size of Europe. In contrast, the total land mass is about equal to the area of Denmark. The total population of the member countries of the USP region is approximately 1.4 million, with national populations ranging from 1,300 in Niue to more than 800,000 in Fiji.

There is great diversity in economic, social, cultural and political structures. The challenges of providing education in the multicultural and multilingual environment in which the USP consortium of countries operates, spanning five time zones and two different days, are very real. The university's distance and flexible learning programme, which commenced 40 years ago, has contributed immensely to tertiary education for the many who might otherwise never have had the opportunity to obtain tertiary qualifications. The possibilities of mobile technologies are thus immense in this academic environment.

The university, established in 1968, is one of two regional universities in the world, the other being the University of the West Indies. The USP is regarded as the leader in the Pacific Region in the provision of tertiary education. The university is owned and managed by the 12 Pacific countries of the Cook Islands, Fiji, Kiribati, Marshall Islands, Nauru, Niue, Samoa, Solomon Islands, Tokelau, Tonga, Tuvalu and Vanuatu. It has a physical presence in each country in the form of a campus, which includes a library. In some countries there are centres on the outer islands. Examples of these are Vava'u and Ha'apai in Tonga and Santo and Tanna in Vanuatu.

The university's administrative structure is very similar to that of most other universities. It is governed by a University Council which includes representatives of each member country and is headed by a vice-chancellor. There are three faculties (Arts and Law; Science, Technology and the Environment; and Business and Economics) led by deans and associated support structures. The library is a support service reporting to the deputy vice-chancellor.

The student population stands at 22,000, registered through face-to-face and distance programmes. Approximately 44% of the full time equivalents (FTE) are external students who study in distance and flexible mode, using self-instruction, print packages, local tutorial support, audio by satellite, visual by video conferencing and recorded lectures through USPNET (USP Educational Telecommunications Network) using class shares and Moodle as the discussion board. With its strategic position and facilities, the university attracts eminent scholars and staff from all over the world.

The USP Library

The USP Library is a network comprising 21 libraries with the main library located at the Laucala Campus in Suva, Fiji and libraries in each of the member countries. The Laucala campus library's 850,000 volumes are managed using SPYDUS (CIVICA Australia) whilst Athena is the library management system (LMS) of choice for the campus libraries. All collections are accessible via an online catalogue.

The main library provides financial and technical support for collection development at the campus libraries, while operations are the responsibility of campus library staff. Of all the campuses, only four are managed by professionally qualified staff – Laucala (Fiji), Alafua (Samoa), Emalus (Vanuatu) and the Solomon Islands. Others are managed by a range of personnel at various stages of acquiring vocational and para-professional qualifications, including one or two campuses without any librarians, which results in its own set of challenges. There are currently no services offered using mobile technology.

In 2008 the Library was part of a university-wide quality audit, the first of its kind in its 40-year history. The Library was the only section in the university to receive a commendation from the Audit Committee, comprised of representatives of the Australian and New Zealand Universities' quality agencies. This was confirmation for the Library that its developments compare favourably with international trends and benchmarks. The interest in attending the International M-Libraries Conference is closely linked to the Library's efforts to respond to the needs of users by making access to services and information as effortless as possible, using ICT, and expanding into mobile technology. Information on the university can be found at www.usp.ac.fj and on its library services and operations at www.library.usp.ac.fj.

The survey, findings and analysis

Just as it is good practice to obtain the reactions of users to the services offered by an organization, it is also as important to obtain the views of users on potential new services. The findings are indicative of the possibilities for learning and teaching. A survey by questionnaire was conducted to obtain a bird's-eye view of users' reactions to the possible introduction of mobile technology into library services. The purpose of the survey was threefold:

1 To obtain information on the range, type and mobile devices owned by library users.
2 To ascertain the number (percentage) of library users with mobile devices.
3 To obtain the views of library users on the use of mobile devices for online learning.

Questionnaires were distributed by library staff at the three campuses of Alafua (Samoa), Emalus (Vanuatu) and Laucala (Fiji). The majority of the target audience were students. A total of 295 individuals participated in the survey, represented by 39 from Alafua, 44 from Emalus and 212 from Laucala Campus. The FTEs for the campuses are 233, 657 and 7530 respectively.

Findings and analysis

The survey found that 94% of participants owned hand-held devices, 97% of these being mobile phones. The 3% of device owners who owned more sophisticated devices such as iPhones or PDAs (personal digital assistants) were more likely to be staff or in-service mature students. The low number of PDAs and iPhones is not surprising, since the minimum cost of these items is in the thousands of dollars, as compared to mobile phones with voice and SMS capabilities, which can be purchased for as little as F$15 (US$7.50).

The survey also found a wide range of brands of mobile phones, the most common being Nokia (44%). Others included Motorola, Alcatel, Samsung, Sony, Sagem, Coral, Digicel, Blackberry, Vodafone, LG, Siemens, Sharp, Canyon and Nile. Since 57% of respondents had voice and SMS features on their phones, this was an indication to the Library of the range of mobile services that it might offer at the outset that would reach the highest number of users.

Users indicated a preference for two-way communication (63%) involving interaction *between* library and user, as opposed to the one-way communication (33%) *from* library *to* user in the form of overdue notices, reservations, library hours, library news, etc. The survey also found that many owned mobile technology that could not support two-way communication. Taking these findings into consideration, the Library will need to consider how it will address these issues. At this point, different levels of services are a consideration.

It was interesting to note from the findings (Table 3.1) that even though the highest percentage of users had only voice and SMS capabilities on their mobile phones, 16% were not using what could be considered basic features. One possible explanation is that the mobile phone is being used as a traditional telephone. A similar finding is noted for the voice, SMS and internet combination, at 24%.

Table 3.1 USP mobile technology survey findings		
Network services	**Available (%)**	**Not being used (%)**
Voice and SMS	57	16
Voice, SMS and e-mail	12	13
Voice, SMS and internet	10	24
All of the above plus MP3	23	n/a
Any of the above plus MP3	7	9
None of the above but MP3	8	14

The main service providers are Vodafone and Digicel, with Vodafone (60%) found to be the majority provider for survey participants, followed by Digicel (36%) and Inkk (5%), a subsidiary of Vodafone.

Although Vodafone has not worked with a library on the provision of library services using mobile phones, the company responded positively. To confirm their interest, Vodafone gave the example of SMS Blood service, which is an agreement with the Blood Bank of the Fiji Ministry of Health that involves the mining of information on registered blood donors who can be contacted to provide blood when needed.

One success story includes a recent case where a cardiac patient who had been operated on by a visiting Australian cardiac team began to bleed late one evening. There was no B+ blood in the hospital's blood bank. Vodafone responded to a request from the cardiac team and sent out a broadcast to B+ donors at midnight that night. Within an hour, three donors had responded, saving the patient's life, as was reported in the *Fiji Times* two days later. Although library services may not always be viewed as a matter of life or death, support by Vodafone for library services using mobile phones is evident. The same was stated by Digicel.

New services are never without their costs. Respondents were asked about their willingness to pay for services and over half (51%) responded negatively. This question was followed by possible reasons for their responses. The most common barriers cited were both linked to costs, the first being *credit not always available* (40%), followed closely by the *prohibitive costs of mobile devices and mobile services from providers* (39%).

One-way communication is thus possible with users' current mobile telephone capacity. Although two-way communication is the preferred option, there are financial implications in terms of upgrading the technology owned by users.

In addition to information on hand-held devices, respondents were also asked if they owned other devices such as laptops, with 51% responding that they owned laptops and netbooks.

The message received from the three campuses surveyed is that the Library has the green light to pursue developments using mobile technology for one-way communication at the outset. The Library offers an information literacy programme for face-to-face and distance learners and mobile technology offers many possibilities for enhancing lifelong learning.

Technology options

The views of the Library Systems section were also sought. Following discussions with the USP Information Technology Section (ITS), the Library learned that ITS is using a GSM (Global System for Mobile Communications) modem with a SIM to send alerts from the ITS monitoring system. Therefore, should the Library decide to implement mobile technology communications, the GSM modem would be a starting point for one-way communication. The way in which the SPYDUS LMS messaging architecture works is that SPYDUS sends an e-mail to a Vodafone e-mail address, e.g. 9246789@vodafone.com.fj, which sends it on to an SMS converter that will then send it to the mobile phone of the library user as an SMS. There are also advantages in terms of charges on a monthly basis, as opposed to individual SMS costs. The Library has yet to confirm whether CIVICA will be able to make SPYDUS function with a GSM modem. In terms of costs, and taking into account the recent 20% devaluation of the Fiji dollar, the possibilities of going with primary providers like Vodafone and Digicel would be ideal, depending on what solutions they can offer. For the Library, this will involve in-depth discussions with all parties, including USP Library, the university's ITS, CIVICA, Vodafone and Digicel on management, technical and financial issues.

Coinciding with the survey was the presence of a staff member from CIVICA, (the company that supports the SPYDUS LMS), who was asked about the capabilities of SPYDUS to support mobile technology developments. SPYDUS has the capability in version 8.4.6 to send reservations and overdue notices by SMS, which is very popular in Australia, with some very positive feedback being received from customers.

The same version also enables access to the catalogue via hand-held devices, and users are able to search the OPAC, reserve books, renew books and ask questions. This has also been well received. A further development is the receipt of selective dissemination of information via RSS feeds.

Before mobile technology services can be implemented at the Laucala campus, an upgrade of SPYDUS from version 8.3 to 8.4.6 is needed. For the Alafua and Emalus campuses, decisions remain to be made on an LMS to replace Athena, which neither has the capability for mobile technology nor is any longer supported by a vendor. With the management, cost and technical implications associated with this development, the challenges are very real.

Challenges for the library

The findings of the survey have provided the Library with information on what it needs to do in order to offer cutting edge services to its users that will enhance their learning through access to information in a timely and convenient manner. The possibilities for mobile technology applications in the Library are endless and for now we note the following challenges which include:

- upgrading the LMS at Laucala to support mobile technology, bearing in mind the cost and technical implications
- identifying a new LMS to replace the Athena system at campuses and centres, preferably open-source, which is sophisticated enough to support mobile technology services; again this has technical and financial implications
- ensuring that the technical infrastructure required to support such an endeavour is in place at all campuses of the university, the Library and USP ITS and service providers
- ensuring parity and equity of access to services in terms of costs to the individual, noting that connectivity in the region is expensive, since the majority of students are private students
- a revision of Library policies governing the use of mobile technology in the library and monitoring these vis-à-vis Library regulations that were put in place to provide a 'quiet study' environment
- other unanticipated issues that will arise as mobile technology applications are investigated in greater depth.

Conclusion

Mobile technologies provide libraries with endless opportunities for learning. It is acknowledged that there will be some trying times, but there is no library that is without difficulties. The USP Library's Strategic Plan for the period 2010–12 supports mobile technology development learning and teaching. The aim is that by the time of the Third International M-Libraries Conference, the University of the South Pacific Library will be offering one-way services using mobile technology at one of its campuses, if not more, proving that the challenges listed are not insurmountable.

Acknowledgements

I wish to acknowledge the assistance of librarians, Angela Jowitt (Alafua) and Margaret Austrai-Kailo (Emalus), systems librarian Chris Hammond-Thrasher (Laucala), systems analyst programmer Vikash Gounder (Laucala) and secretary Cherie Fonmoa-Kean for their assistance. Last but not least, my thanks to the university and the Library for conference funding.

4

M-Library in an m-University: changing models in the Open University of Catalonia

Dora Pérez and Pep Torn

Introduction

One of the early images used by the Universitat Oberta de Catalunya (Open University of Catalonia, UOC) to explain its educational model when it was created in 1995 was that of a young man seated under a tree, working on his laptop computer. The image was not so specific as to be able to determine whether the student was working in his garden, in a city park or in the middle of a mountain range, but it was evident that he was a student 'on the move'.

The spirit of the UOC includes mobility. The appearance of mobile technologies – *m-technologies*; of newer and ever more *mobile* devices than the first laptop computers of the 1990s – *m-devices*; and of suitable formats for consultation using these devices – *m-formats* – is a panorama of continuity for universities such as ours that made a commitment to ICT from the start, as the basis of their project, and to virtuality as a medium.

Virtuality today, however, is a widely implemented model in the university world, and evidently the Spanish scene has ceased to be the exclusive domain of the distance university. Many degrees, courses or subjects are offered in e-learning or blended learning formats by the traditional universities, and this means, among other things, that the collections and services of the libraries of these universities have adapted this model to a greater or lesser extent. What is the case, then, for mobile devices, the latest devices to appear on the virtual scene? What is the implication for the UOC and its library regarding the adaptation of collections and services for this type of device?

In this chapter, we look at the Spanish situation regarding m-technologies on the basis of surveys conducted between April and May 2009 with the directors of Spanish university libraries. At the same time, we show the path taken and the present situation of the UOC, the university where we work, in terms of the use of m-technologies, m-formats and m-devices.

M-libraries in Spain

In Spain there are at present 73 universities. The libraries of 71 universities, of which 70% are publicly owned and 30% privately owned, and the libraries of the state research centre, Consejo Superior de Investigaciones Ciéntificas (CSIC), make up the REBIUN network (Red de Bibliotecas Universitarias Españolas, or Spanish University and Scientific Libraries Network), a university library network that shares – with greater or lesser involvement by each participant – the development of a common strategic plan, collaborative work on projects of shared interest, etc.

A study of these libraries was carried out to discover the level of development of access to content and services using mobile devices (Zawacki-Richter, 2007). The study consisted of a survey of all the libraries with the aim of finding out the level of implementation of m-technologies. The survey comprised just three questions, so as to obtain a good percentage of response and gain a broad view of the use of these services and of the current level of interest of libraries in new developments. The questions were:

1 Does your library have or is it planning to have any type of service that can be accessed using any mobile device?
2 What type of service accessible from mobile devices does your library offer?
3 What devices are used to access these services?

Of the 71 libraries, 63 replied to the survey, i.e. almost 90%, producing results that are highly indicative of the Spanish situation (Figure 4.1).

Of the 63 libraries that replied, 18 provide access to services via mobile devices, 45 do not have such services and 18 are planning them. It should be noted that of the 18 libraries that are planning services, at the time of the survey 12 had no service of this type in operation and 6 had some service in operation but were planning to offer new services of this type to their users (Figure 4.2).

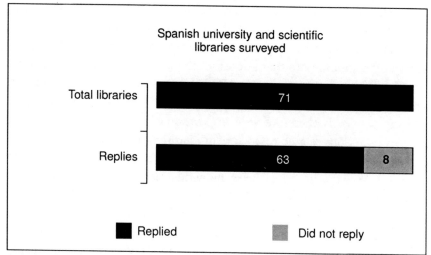

Figure 4.1 REBIUN libraries surveyed

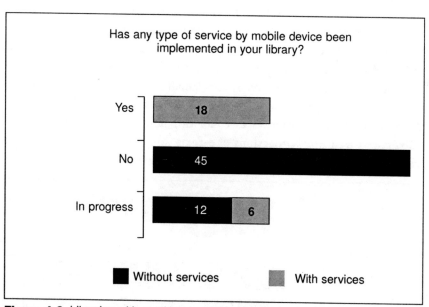

Figure 4.2 Libraries with or without m-technologies implemented

M-library services

In light of the survey, four categories of service have been identified: document lending, general information, OPAC (Online Public Access Catalogue) related service and e-text oriented service.

Of the 18 libraries that have implemented services, 14 have services related to the lending of physical documents: renewals, overdue claims, reservations, etc. A total of 13 more libraries are planning this type of service. We should note that almost all the libraries surveyed are on-site libraries, with the exception of the UOC (Universitat Oberta de Catalunya) and the UNED (National Distance Education University), and that the lending service is therefore very important for all of them. The most numerous actions related to document lending are overdue claims (eight libraries), followed by reservations (seven libraries) and renewals (five libraries). Very few libraries (three) allow their users to request documents on loan via mobile devices (Figure 4.3).

In numerical terms, the lending service is followed in importance by the information service (ten libraries provide an information service and five are implementing it), which includes different types of information: library opening hours, information about the lending service, new items, etc. We should note here a service offered by the UOC which provides its users with thematic news services of varying frequency using various devices.

Access to OPACs via mobiles and PDAs also has a leading position among those services using mobile devices. Six libraries offer this service and five are working towards offering it. Finally, just one library offers access to e-texts, although three more are about to start up this service. It should be remembered that we are evaluating libraries, and there are several universities that offer collections of e-texts and access to e-book devices from other university units or departments apart from the library.

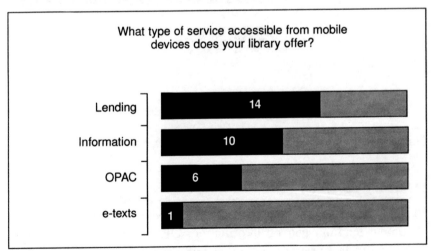

Figure 4.3 Types of mobile service implemented

Devices and technologies in m-libraries

The devices used to access the services described above are divided among the following technologies: mobile telephone in general, Short Message Service (SMS) and personal digital assistant (PDA) (Figure 4.4). The SMS service is the one most widely used for renewals, reservations and overdue claims in document lending services, while mobile telephones, followed by PDAs, are the most widely used to access the OPAC and information services.

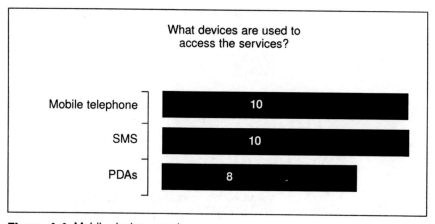

Figure 4.4 Mobile devices used

Planned services

A total of 18 libraries are planning services via mobile devices. The majority (12) do not presently have services but are working towards implementing some. The remaining six already have services in operation but are planning to extend their range.

Almost all the libraries provided information about the projects that they are planning, and the vast majority are tending towards services related to document lending (13), some are considering offering information services and access to the OPAC (5), and three are preparing e-book collections and devices (Figure 4.5). Three libraries are also planning to adapt their whole website to make it accessible using different devices.

The devices or technologies for access to planned services are based primarily on SMS and mobile telephone in the great majority of projects (Figure 4.6). A trend is starting in the use of e-books to provide access to e-texts and make them available to users via the lending system.

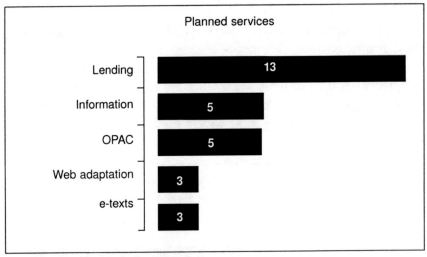

Figure 4.5 Planned m-services

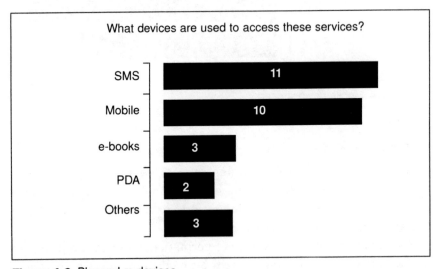

Figure 4.6 Planned m-devices

Discussion of the Spanish situation

The conclusions that can be drawn from the study show that although it is not the main concern of libraries, many of them are interested in providing their users with access to the library without having to go there in person.

At a European-level professional meeting in June 2009 (attended by Spanish library directors), participants were asked to identify the trends of the coming two or three years in their libraries, and ten were identified. When they were asked to reduce these further to just three trends that they saw as key, among those three was the application of mobile technologies.

To return to the survey, 29% of the libraries surveyed have already started down this route, and a further 19% will probably be added to this figure by late 2009 or early 2010, to make 48% of Spanish university libraries soon to be offering services accessible from mobile devices, even though they are on-site libraries. If progress is made in the downloading of e-texts for use by students and lecturers, this service may be adopted by more libraries, as there are presently many libraries that are offering or about to offer an e-book lending service, although without a sufficient amount of content.

Experience tells us that, at a Spanish level, common undertakings produce better results than individual developments. Work in environments such as consortia or professional networks has enabled us to achieve levels of adaptation to ICT, of professionalization of human resources and of access to collections far greater than before. It is interesting to find that commitment to the creation of m-libraries is an aim shared by many Spanish universities. This means that when proposing projects, defining objectives and sharing experiences it will prove to be easier to gear actions towards successful outcomes.

The Universitat Oberta de Catalunya, the library and m-technologies

Aside from the transfer of research to society that is performed by the UOC Internet Interdisciplinary Institute (IN3), a significant part of the innovations made at the university are tested and developed for application within the university. The educational, communication, service – and so on – model of a virtual university is being constantly updated, and the UOC is a test bed for all of the technological innovations that can be applied in the university environment to improve it in all aspects. It is in this environment that the UOC made a commitment in the 2000s to conduct research into the adoption of mobile technology to offer services to the virtual community. The first developments were through SMS and WAP (Wireless Application Protocol).

The OPAC and WAP technology

In February 2000, the UOC became the first Spanish university to offer Virtual Campus content using WAP technology. In light of the potential of this new technology and forecasts for its growth, it was felt that it opened up a new opportunity for libraries to offer their services and resources via these new devices (Serrano, 2000). The structure of the WAP server was implemented on Windows NT using Nokia WAP Server. The services that were implemented with this technology throughout the university were:

- procedures with the Secretariat
- academic calendar
- information about support centres
- mailbox
- new features on the Virtual Campus
- access to the library OPAC.

For access to the library OPAC, which at that time used the VTLS software on ORACLE, the Virtual Web Gateway was duplicated to enable a web listener by WAP. This PERL web gateway was recoded to adapt the HTML output to the VML language used by WAP devices, and the number of results was limited to 5 by 5 rather than 20 because of the size of the devices' screens. Accessing from the main menu, the user selected the search option with the usual searching possibilities: author, title, subject area, subject. The editor that was incorporated into the device assisted the entry of the search parameters and sent the information to the server and, after a few seconds, the search results were received. It seemed at that time that later WAP generations would allow new services to be offered, so the Library prepared to develop a system of reserving loan documents and access to the e-documents available in the UOC catalogue, with information about their location in the central warehouse or at the different UOC support centres across the region.

GPS, SDI and reservations

Another line of work was that of informing the user of the location of the nearest document to the place where the reservation had been made, using GPS, and the possibility of receiving the different 'push' services that had been implemented in the Library at that time: SID (Selective Information

Diffusion), the different themed news services, etc. Another development in this vein was in the use of SMS technology. Taking advantage of the development work done by the university to send out marks by SMS, work was carried out to adapt this system for sending document lending transactions, specifically to send confirmations of loan reservations and overdue loans claims.

The results and conclusions that were reached through the development of these new technologies were that internet access by mobile telephone would allow users access to information without depending on computers and terrestrial communications. This implied a new opportunity for libraries to offer their services on and adapt them to this new environment. Within the context of the university, the UOC Library offered the services, developed over a number of years, to the virtual community as complementary or alternative to access via computer.

Because of other priorities, the university did not pursue the development of mobile technologies, and so the Library also abandoned its developments, which would not be feasible if the university did not make progress in these technologies into the Virtual Campus. At the same time, the Libraries Coordination Unit of the Higher Science Research Council (CSIC) collaborated on the project World Wide Web Information Server of the Pacific Science Commission (II): Digital Dissemination Systems of Cultural Heritage (TIC 2000-0168-P4-04), which investigated access to an information system using WAP mobile telephony.

M-technologies in the UOC

The present situation is rather complex. With the widespread take-up by the public of mobile telephones as a means of communication and the range of services on offer, the proposed projects of the UOC are very extensive. In terms of its extensive use, the university's most notable mobile application is the sending of final course marks to users' telephones individually. This system has been in use since the 2000–2001 academic year. However, the library is not a participant in this project.

Another notable project is one that has been carried out during the current academic year with a number of postgraduate courses or individual subjects, where all the students enrolled have been provided with an e-book device (iRex iLiad) that contains all the teaching modules for the subjects that the student is studying. This project has been carried out on a pilot basis and the students' assessment of it has been uneven.

The specific issue of digital literacy has been analysed in depth in order to determine levels of satisfaction and the device's suitability to users' needs. The students in the analysis are all in humanities courses, divided into psychology, educational psychology and other courses. This is also a highly expert population within the university environment, with 90% of them studying for their second degree at the UOC and all having studied at the university for two or more terms. It is a group with proven skills in the use of office technology (all UOC students have to prove this through mandatory instruction before enrolling on the degree course). The average age of the users to whom the device was provided is 33, and they are primarily (>60%) women.

Contrary to what one might have thought, in view of the profile of the sample, the first surprise for the Educational Technology team when analysing the project was that, generally speaking, no one knew the operation, the general utility and, evidently, the technical operation of the devices. Of the users analysed, 100% resorted to the internet to find out how to make the equipment work. Especially remarkable is the fact that 100% of the users have continued to use all of the other usual materials and tools at the UOC to complete the courses in question, such that we can say that they have seen the device as a non-essential support tool.

We believe that this should be understood in the specific context of the UOC, where the students who have worked with the virtual campus for a number of terms know the work methodology and have incorporated it into their way of studying. Thus, to use a device like the iRex iLiad is really an alternative rather than a main option – more so when the course materials can be used without using the mobile device. The two aspects that the students saw as most beneficial were the possibility of the device's not only allowing the teaching materials to be read, but also storing extra information generated by them, and the ease of reading on screen, for which they considered the device to be more useful than a laptop computer.

The evaluation by the UOC department that conducted the pilot trial is generally positive. The most remarkable issues that Educational Technology staff highlighted on completion of the trial were, first, the modernity and improvement of the institution's image. On a practical level, they stressed the ease of sending materials, as they were not reliant on messaging services to do so. The materials can be sent all over the Virtual Campus, and the user has the option of downloading them on e-reader rather than paper; this aids simultaneity of distribution. Finally, we should also highlight the ecological aspect implied by the saving of paper that is achieved by using 'electronic ink' devices. Contrast this with the traditional scenario of the University having no option but to print and distribute all course materials in paper format.

At the end of the course, the students were able to buy the devices, paying 50% of the original price, and the result was that 25% of the students decided to buy them for personal use. The Library did not participate directly in this project, although 40% of the devices returned by the students have been provided to the Library for addition to its inventory of equipment and for loaning out.

Current m-technologies and m-projects in the library

The involvement of the library during this period, which we could call 'second era', is still in rather early stages. After analysis of the 'first era' of developments and commitments under the WAP system in the first decade of the 21st century, there is greater caution before committing to specific technologies. In September 2009 we will be starting the 2009–10 academic year with a range of 18 e-book devices in the UOC libraries. Of these, 15 are the returns of students on the postgraduate Open-Source Software course, the iRex iLiad devices, which allow consultation of the teaching materials for many UOC courses. The other three devices are the iRex *DR 1000*, which has an A4-size screen and which allows all the teaching materials for all UOC courses to be read.

The commitment of the Library is to offer these devices on loan in a way very similar to that in which printed books are offered, as the format should not determine the content of what is on offer, we treat the devices on the basis of their content, i.e. as if they were another book. The content that we make available to users for consultation via e-books is included on a library web page and ranges from teaching materials created by the UOC for all its various courses and programmes, to material from the Library's digital collection. The Library's Documentary Resources team has conducted an exhaustive analysis of the documentary sources purchased by the Library in order to identify which ones can be consulted on the iRex. Thus, users can have access to a significant collection of materials to read on the device.

The other projects on which the Library is working are University-wide and regional technological solutions, with which all the Library's resources are managed. The libraries in the Consorci de Biblioteques Universitàries de Catalunya, or Catalan University Library Association (CBUC) have jointly acquired the Innovative Millennium, SFX and Metalib documentary management software, and it is in this environment that we have worked the hardest to apply software solutions to the optimal level.

Conclusions

In light of the proliferation of mobile technologies among our users, the m-library concept is something to which we are strongly committed, but in which we have to take firmer and more secure steps than those we have taken up to now. Our first experience in the commitment to WAP technology told us that we need to combine with the other technology departments at the UOC, with a view to making a commitment to technologies with which to work on areas of the m-library that we wish to develop.

At a markedly technological university, issues relating to the implementation of new technologies are of great importance. The UOC is supported by a complex structure in which the technology support teams used by all the departments are key, such that part of the technological knowledge resides not within the Library itself, but in the teams that provide support to the Library from other areas. There are primarily two departments responsible for ICT support for the Library: Applications and Processes, and Educational Technology, and the steps that we take from now on in committing to mobile technologies have to be agreed upon with these two departments.

The strongest commitment made by the Library in recent months – the acquisition of the 18 e-books readers – was made on the basis of joint work with the Educational Technology department. This way, we ensured the support of the institution in the event of any problems with the devices, with training needs or with needs to share knowledge.

With regard to the m-technology trends with which the Library considers it is suitable to work, we should say that we have our reservations in terms of devices which are presently hardly used among the public. Although it is true to say that the UOC is seeking to be a trendsetter in Catalonia and Spain in aspects of innovation, it is also true to say that users set the agenda, in part, when it comes to defining the communication tools to which we should be making a commitment.

We need to have good knowledge of the environment in which we will be working with future trends, and the Catalan and Spanish case is highly specific. An EU study conducted in 2006 placed Spain above the European average for mobile telephone penetration, specifically with 94% of users having a mobile telephone, whereas the figure for the EU as a whole stands at 92.8% (Figure 4.7).

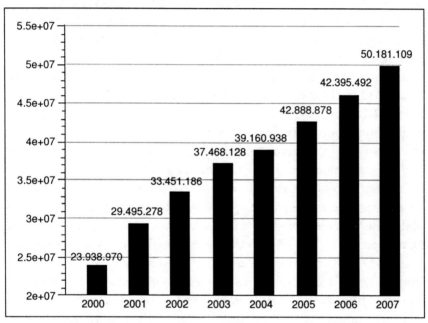

Figure 4.7 Growth of the number of mobile telephone lines in Spain between 2000 and 2007
Source: Red.es Observatory

Although anticipating user needs is important, sticking our neck out in thinking that users will choose on a large scale to buy mobile devices other than telephones is rather risky. In a recent conversation between David W. Lewis (Dean of the IUPUI University Library and Indiana University) and a number of directors of Catalan university libraries, Lewis offered the following question for reflection: what is the age of the users that we have seen holding electronic ink devices? Generally speaking, we can say that they are all rather older, say, over 30. By contrast, there is great diversity in the ages of the people whom we know to use a mobile telephone on a daily basis. In this environment, we at the UOC Library see mobile telephones as the technology of the future to which the Library should devote a large part of its efforts, and use of electronic ink devices should be approached more cautiously.

Although we regard our library, like every other library in the world, as an area with a certain degree of autonomy in some issues, the Library's wish is for shared work with all areas of the university, and the whole university is involved in permanent technological change, in such a way that the UOC should develop adoption of m-technologies as a whole, and the Library in particular. A virtual university such as the UOC should also be an m-university and have, among other things, an m-library.

Besides this, and in the same way that we feel that shared work between libraries all over Spain can be enriching when defining the model(s) of m-libraries on which to work, the UOC Library, as a member of the CBUC, is immersed in an environment of various documentation management products. These products are primarily and presently software related to or compatible with Millennium, SFX and Metalib. It is in these environments that we are directing our efforts in order to adapt the present virtual library to become a virtual library for mobile devices. At present, we are considering the possibility of incorporating the Millennium AIRPAC module (which allows consultation of the OPAC by mobiles and PDA) and SMS Notice module (which allows alerts to be sent relating to the circulation of collections) into the library catalogue. We trust that the Innovative and Ex Libris developments will also take this direction, because the UOC Library, urged on by the commitment of its users, has started out on a road towards mobility that will stop only if the users make such a demand.

Acknowledgements

We would like to sincerely thank the Library Directors of REBIUN for their support and collaboration in the questionnaire.

References

Serrano, J. (2000) Acceso a la Biblioteca de la UOC por medio de la telefonía Móvil, *I Jornadas de Bibliotecas Digitales*, Valladolid, 6–7 noviembre 2000, http://biblioteca.uoc.edu/

Zawacki-Richter, O., Brown, T., and Delport, R. (2007) Factors that May Contribute to the Establishment of Mobile Learning in Institutions – Results from a Survey, *International Journal: Interactive Mobile Technologies (iJIM)*, 1 (1), 40–4.

Further reading

Ally, M. et al. (2008) Use of a Mobile Digital Library for Mobile Learning. In: Needham, G. and Ally, M. (eds), *M-Libraries: libraries on the move to provide virtual access*, Facet, 217–27.

Arroyo-Vázquez, N. (2009) Web móvil y bibliotecas, *El profesional de la información*, (marzo–abril), 18 (2), 129–36, DOI: 10.3145/epi.2009.mar.02

Johnson, L. F., Levine, A. and Smith, R. S. (2009) *Horizon Report*, The New Media Consortium,
www.nmc.org/news/nmc/horizon-2009-translations

Keegan, D. (2003) The Future of Learning: from elearning to mlearning,
http://learning.ericsson.net/mlearning2/project_one/book [accessed 13 September 2006].

Keegan, D. (2005) *The Incorporation of Mobile Learning into Mainstream Education and Training*. Paper presented at the World Conference on Mobile Learning, Cape Town.

López-de-la-Fuente, J.-M. (2009) Merkur: herramienta de transcodificación parametrizada de contenidos, *El profesional de la información*, 18 (2), (marzo-abril), 218–22, DOI: 10.3145/epi.2009.mar.12

Quin, C. (2000) mLearning: mobile, wireless, in-your-pocket learning, *LiNE Zine*, Fall 2002,
www.linezine.com/2.1/features/cqmmwiyp.htm

Sheikh, H., Eales, S. and Rico, M. (2008) Open Library in Your Pocket: services to meet the needs of on- and off-campus users. In: Needham, G. and Ally, M. (eds), *M-Libraries: libraries on the move to provide virtual access*, Facet, 187–96.

Tin, T., Sheikh, H. and Elliott, C. (2008) Designing and Developing e-learning Content for Mobile Platforms: a collaboration between Athabasca University and Open University. In: Needham, G. and Ally, M. (eds), *M-Libraries: libraries on the move to provide virtual access*, Facet, 173–83.

Zawacki-Richter, O. (2005) Online Faculty Support and Education Innovation – A Case Study, *European Journal of Open and Distance Learning (EURODL)*, Vol. 1, www.eurodl.org/materials/contrib/2005/Zawacki_Richter.htm

Zawacki-Richter, O., Brown, T. and Delport, R. (2009) Mobile Learning: from single project status into the mainstream?, *European Journal of Open, Distance and E-Learning*, *EURODOL,*
www.eurodol.org/materials/contrib/2009/RichterBrown_Delport.htm.

5

Piloting mobile services at University of Houston Libraries

Karen A. Coombs, Veronica Arellano, Miranda Bennett, Robin Dasler and Rachel Vacek

Introduction: background of the pilot project

In the fall of 2008, a small group of librarians at the University of Houston (UH) Libraries embarked on a pilot project to develop, deploy and evaluate mobile services for library users. The overall objective of the project was to discover how mobile devices could be used to enhance the services currently provided by the library and, concurrently, to investigate mobile technology efforts at other libraries.

As demonstrated by the 2009 conference programmes for two major library associations, librarians' interest in the use of mobile devices is increasing. At the Association of College and Research Libraries 14th National Conference, general presentations on mobile devices were offered, as well as programmes on using mobile devices to provide roving reference and serve the on-the-go mobile learner. Attendees at July's American Library Association Annual Conference in Chicago learned about the WorldCat Mobile pilot project, mobile devices 'on the road to the future', and the public policy considerations linked to libraries' use of mobile devices.

Around the country, librarians are experimenting with providing library services designed for mobile device users, including audio tours, text message alerts, MP3 tutorials, and mobile reference. A good summary of these and other projects can be found in Kroski (2008). The impetus for creating such new services is the clear trend toward ever greater mobile device usage among library patrons. According to the ECAR Study of Undergraduate Students and Information Technology, 2008, 'One of the most significant trends we report this year is the continuing "mobilization" of the student body', as internet-enabled mobile phones become more and more common on campus. Likewise, the 2009 Horizon Report put 'Mobiles' in the category 'Time-to-Adoption: One Year or Less' and noted that '[t]he rapid pace of innovation in this arena continues to increase the potential of these little devices, challenging our ideas of how they should be used and presenting additional options with each new generation of mobiles' (Johnson et al., 2009). Clearly, mobile services must be a priority for librarians who want to meet patrons' needs in a rapidly changing technological environment.

Notably, nearly all library-related mobile projects have focused on developing services for patrons: mobile-friendly catalogue interfaces, mobile chat reference, information-literacy podcasts, etc. To meet the needs of library users most effectively, however, librarians must themselves work with mobile devices so that they are familiar with their features and aware of the patrons' perspective. In addition, the portability and convenience of mobile devices offer distinct advantages to librarians in their daily work as they go beyond their offices to provide expert assistance around a campus or community, or wherever their responsibilities take them.

As a result, the members of the mobile device pilot project team decided to focus on providing librarians with mobile devices in order to discover how these devices could enhance their work. Essentially we wanted to investigate three questions:

- Could mobile technologies help librarians to work more effectively?
- Could mobile technologies improve services offered by library liaisons?
- Do mobile technologies change the way in which librarians work?

Implementing the pilot project

The pilot project was designed to take place from the fall of 2008 to the spring of 2009. Since the focus of our pilot was on librarians working with mobile devices, the UH Libraries purchased eight iPod Touches. Four of them were assigned to members of the mobile device project team, which included a member of the Web Services department, a member of Collection Services, and two subject librarians (one a liaison to science departments and the other to the social sciences). One iPod Touch was assigned to the liaison librarian for Business and Economics, and the remaining three devices were made available to all UH librarians for checking out. The mobile device project team developed a set of policies and procedures for use of the iPod Touches, as well as basic support documentation for the use of the devices. This information was made available on the UH Libraries intranet.

Each iPod Touch contained a basic set of tools, including the basic iPod apps (Safari, Calendar, Mail, Notes), Stanza Ebook reader, AIM chat, Briefcase Lite, Wordpress and Instapaper. Participants were allowed to install additional free apps from iTunes or could submit a request to Web Services to purchase an application.

In order to garner librarians' interest in the mobile device project, team members held an iPod Touch show-and-tell at two different meetings of UH subject librarians, as well as two iPod Touch training sessions open to all library staff. At each of these meetings and training sessions library staff were encouraged to use the iPod Touch and familiarize themselves with the accompanying iPod Touch policies and procedures. Additionally, e-mails were sent to librarians describing the project and making them aware of the availability of the iPods for their professional use.

During the pilot project, team members with assigned iPod Touches kept a record of their experiences on a collaborative blog (http://weblogs.lib.uh.edu/mobile/). All library staff who checked out an iPod Touch for on-campus or conference use were encouraged to post their user experiences to the blog as well.

In order to assess the mobile device pilot project, team members developed an online survey using SurveyMonkey and distributed it to librarians via a library listserv. The survey was designed to elicit feedback from librarians who used an iPod Touch as well as from those who chose not to do so. All survey respondents were asked if they would be willing to be interviewed regarding their experiences in the project. A total of 27 people responded to the survey, 14 of whom volunteered to be interviewed.

What we learned
Pilot project participants

After examining the survey data, we learned that 44% of our survey respondents had used an iPod Touch at least once in varying locations:

within the library, elsewhere on campus or at home. Subject librarians took the iPod Touch to meetings with faculty for convenient access to their e-calendars and the library catalogue. Other librarians and library staff took the iPod Touch to on-campus orientation sessions and fairs for new students and faculty to demonstrate use of the UH Libraries website and e-resources.

By far the most extensive use of the iPod Touch took place at professional conferences (all but one of the respondents reported using the device at such an event). Librarians used the iPod Touch to stay in touch with students, faculty and colleagues via e-mail and to chat while out of the office. Yet this mobile device was also used in ways we had not anticipated, including taking notes at conference sessions, keeping track of conference meetings, locating vendor booths at the exhibit hall and creating a list of ISBNs for future book purchases.

Other unexpected uses of the iPod Touch included our Program Director for Collections listening to podcasts related to her collection subject area to keep up with the latest research developments. Our Science and Mathematics librarian discovered that the iPod Touch could also be an effective tool for testing access to electronic resources. Because the iPod Touch is Wi-Fi enabled, it was easy for her to access electronic resources via the campus wireless network, which functions in the same way as for users accessing electronic resources from off-campus locations.

Regardless of their location or intended purpose, library staff primarily used the e-mail, web-browsing, and note-taking applications on the iPod Touch. Fifty-eight percent of users reported using iCal to check and update calendars, 50% reported using the directions/maps application, 33% used a chat application such as AIM or Meebo and 25% used some kind of an RSS aggregator/reader.

Although we anticipated that subject librarians would use the iPod Touch to aid in their library liaison activities (67% reported that they would like to do so), only 25% of our users reported actually using it for this purpose. This may have been due in part to unstable wireless connectivity on campus or an initial lack of comfort with the device.

Overall, pilot project participants were relatively happy with the iPod Touch. Fifty-eight percent of respondents who used it at least once reported that they would use it again. Thirty-three percent would use the iPod Touch again under certain circumstances, including being able to use it for music applications, being able to sync the device with their desktop programs (specifically e-calendar), and having better access to a wireless connection both on and off campus.

Non-iPod Touch users

Data from survey respondents who chose not to participate in the pilot were also informative. The two primary reasons respondents gave for not using an iPod Touch were (1) they didn't know they could check one out and (2) they already owned a mobile device. The third most commonly cited reason for not using the iPod Touch was that respondents felt limited in what they could do with the device, since it did not belong to them. Several other respondents also described not having enough time or incentive to try to incorporate the iPod Touch into their work routine.

Despite the team's best efforts at publicizing the project within the library, there was still a lack of awareness about the pilot amongst some staff. This may have been exacerbated by the arrival of an unexpected storm – Hurricane Ike. Because of mass power outages in the Houston area, there was a break in normal work flow which led to a delay in the start of the project. As a result, our team was forced to limit the number of iPod Touch training and information sessions offered. Better communication about the project, its intent and procedures might have resulted in a greater number of participants.

Video interviews

In addition to data gathered from library staff via a web-based survey, certain survey respondents shared their thoughts about the mobile device pilot project through video interviews conducted by the project team. The intent of the interviews was to enable survey respondents to describe their involvement in the project in greater detail and share thoughts on the relevance of mobile devices to their individual 'brand' of librarianship. Fourteen survey respondents were interviewed; ten had used an iPod Touch and four had not used the device.

The video interviews reinforced a number of key points from the project. Almost everyone who participated in the pilot felt that the iPod Touch was invaluable as a means of remaining professionally connected. Library managers were able to conveniently check in with staff, and subject librarians could answer questions from students and faculty regardless of their location (in the office, out on campus, at conference, etc.). Ironically, although participants wanted a convenient means of connecting to their work, they didn't want a constant out-of-office connection to be mandatory. Our Business and Economics Librarian said it best when she stated that mobile devices change the expectations of library users and colleagues. Although staying professionally connected definitely has its benefits, there is concern expressed by our librarians over how to manage the expectations of 24/7 availability that come from users with mobile devices who know that you have the same technology capabilities.

Communication issues aside, the iPod Touch did help librarians to use their time effectively. One librarian discussed how he was able to check the resumes of potential candidates from the placement centre at the American Library Association (ALA) conference during his free time at the conference.

Despite the popularity of the iPod Touch, our interviews taught us that one size doesn't necessarily fit all. While some librarians felt that the size and portability of the device was excellent, others felt constrained by the small display screen and touch-screen keyboard. Several of these participants voiced a preference for a netbook rather than an iPod Touch. Ubiquitous access was also an issue. Without stable Wi-Fi access, the iPod Touch is practically a paperweight. This was a problem in some buildings on campus, but a much larger issue for librarians at conference, where Wi-Fi was not provided or only available for purchase at conference hotels. As a result, many librarians expressed a preference for an iPhone (with its continuous network connection) and a few indicated that they might be purchasing one for personal, and inevitably, professional use.

The variety and number of mobile devices available on the market raised the issue of device overload. Several technology-savvy librarians who already owned mobile devices questioned how the iPod Touch could fit into their lifestyle. Their ultimate desire was to have the fewest technological devices with the most functionality and flexibility for both professional and personal use. Many different options were suggested (e.g. netbook + iPod, iPhone + laptop, iPod Touch only), but what was never disputed was the need to have access to convenient and useful technological tools.

Next steps

Mobile devices for all

This pilot project established a need for mobile devices amongst UH librarians to aid their professional productivity. Given the expressed preferences librarians had for a variety of mobile devices (iPod Touch, iPhone, netbook, etc.), our project team recommended that librarians be given the opportunity to choose from different library-purchased and owned mobile devices. A proposal was submitted to our library deans and accepted. Beginning in fall 2009, librarians will be able to choose between a netbook or an iPod Touch as a technological supplement to their standard desktop computers. We believe that this new option will give librarians the ability to stay connected to their work while out of the office and allow for more flexibility in meeting the needs of students and faculty. Our team also hopes to use this period to gather more data on how librarians use mobile devices. A review of librarians' experiences with the new mobile technology option will be conducted in one year in order to assess the efficacy of the netbooks and iPod Touches amongst a larger group of librarians.

Improved Wi-Fi access

Since both the netbook and iPod Touch are Wi-Fi dependent, we also recommended conducting a survey of wireless coverage in all areas of the UH libraries. During the pilot project many librarians found low or no Wi-Fi signal in several key staff and student areas within the library. Ideally, these Wi-Fi 'holes' would be systematically discovered, documented and reported to the university's information technology services department. Additional access points would, it was hoped be added to expand the areas in which the netbooks and iPod Touches could be used.

Library apps

One of the most interesting projects that has developed from our initial pilot is the beginning of a collaboration with a computer science professor at UH. In one of his courses he gives graduate and undergraduate students the opportunity to create iPhone and iPod Touch applications that can be used in real-world scenarios. The librarians who used the iPod Touch during our pilot project developed a long wish list of library-related apps that would help to improve the professional functionality of the iPod Touch – apps such as a mobile version of our library's catalogue, a listing of all subject librarians and their contact information, or even just the location of the library and its hours would be great tools for librarians and library users. Utilizing the skill sets of UH computer science students will help to bring components of the library app wish list to fruition.

Conclusion: new directions for the library

Our work on this pilot project has reinforced our belief that the mobile arena is both vast and increasingly important for libraries. E-books on mobile devices, roving reference librarians, SMS reference services or a mobile virtual presence are just a few of the additional mobile projects that sparked our team's interest. All were well beyond the scope of our pilot project, yet all could greatly benefit library services. Therefore, our most pressing recommendation to our library's administrators was that the expansion of mobile library services be given priority as one of the library's areas of emphasis during our new strategic planning process. The importance of mobile devices for librarians was reinforced by our pilot project, which revealed even greater opportunities for 'mobile expansion' within our library.

References

Johnson, L., Levine, A. and Smith, R. (2009) The 2009 Horizon Report, The New Media Consortium,
www.educause.edu/ELI/2009HorizonReport/163616

Kroski, E. (2008) On the Move with the Mobile Web: libraries and mobile technologies, *Library Technology Reports*, 44 (5), 33–8.

Part 2

Technology in m-libraries

6

Evolution of modern library services: the progression into the mobile domain

Damien Meere, Ivan Ganchev, Máirtín Ó'Droma, Mícheál Ó'hAodha and Stanimir Stojanov

Abstract

This chapter describes the main elements of a service architecture needed to support the expansion of existing library-based services into the mobile domain, based on a model proposed as part of the Distributed e-Learning Centre (DeLC) initiative. The enhanced DeLC architecture, detailing the underlying communications infrastructure, along with the various enhanced library-based services, is discussed. The capabilities of these mobile services (m-services) in ensuring greater dissemination and reorganization in relation to the large volumes of administrative information which tertiary education institutions are required to deal with on a daily basis are considered. The utilization of various profiles in order to facilitate a more personalized and contextualized information environment for library users is detailed. Finally, an approach to the development of 'Personal Assistants' (PAs), operating within this multi-agent environment is outlined.

Introduction

The incorporation of mobile technology into academic spheres with a view to 'anywhere/anytime/anyhow' m-learning opportunities has grown exponentially in libraries and campuses worldwide. While the adoption of technologies such as SMS (Short Message Service)/MMS (Multimedia Messaging Service), podcasting, etc. into the realms of education and learning is tremendously exciting, their integration so as to facilitate 'learning on the move' within the library/information environments has not as yet achieved its transformational ideal. This chapter builds on a collaborative research initiative known as the Distributed e-Learning Centre (DeLC) – established between the University of Limerick and the University of Plovdiv – the aim of which is the provision of distance m-learning/m-library facilities to students, educators and information workers. The underlying structural design of the DeLC is dealt with in the first section.

A particular focus in this chapter is the use of podcasting and the promotion and adoption of SMS/MMS. These are essential elements of the mobile wireless communications service infrastructure in delivering m-library services to users/information seekers throughout a university campus. The enhancements facilitated to traditional library services and practices through the incorporation of these technologies are explored in the second section, highlighting how these technologies have previously been adopted and the advantages they can afford to library users and learners in general. Two m-library services, 'Interactive Library Map' and 'Library Catalogue, Loans and Reservations', to be supported by this infrastructure also receive particular attention in this section. The amalgamation of these services in order to facilitate an automated recommendations service is looked into.

The multi-agent service-oriented version of this DeLC, with its technological support for an enhanced mobile environment is examined. Integral to m-services within the library environment are issues of ownership and service complexity management. In the third section, ideas pertaining to the provision of increasingly contextualized services, and the ability to afford system users a great deal more ownership and control over their interactions with the presented services are discussed. Trends towards student 'ownership' of their own learning interactions, and how the capabilities of these mobile processes may ensure better dissemination and reorganization in relation to the large volumes of administrative information which tertiary education institutions are required to deal with on a daily basis are dealt with in detail. The fourth section concludes the chapter.

System architecture

The system architecture proposed here to facilitate library-based m-services stems from the InfoStations paradigm. This infrastructural system concept is based around the idea of facilitating system users with wireless access (via Bluetooth or WiFi connection depending on network constraints) to localized and contextualized services through a distributed network of wireless nodes (InfoStations) situated at key points throughout a campus (e.g. library). Ideas pertaining to this system have been discussed previously in (Ganchev et al., 2008; Ganchev, O'Droma et al., 2008a, 2008b, 2008c). The architecture is organized as a three-tiered structure, incorporating user mobile devices, InfoStations and an InfoStation Centre as illustrated in Figure 6.1.

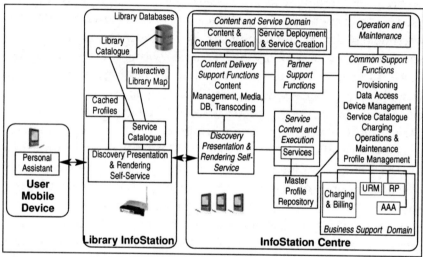

Figure 6.1 InfoStation-based system architecture

The first tier encompasses the user domain, populated by users' mobile devices equipped with intelligent agents acting as Personal Assistants (PAs) to the users. Each PA facilitates an InfoStation with all the information it needs to adapt and customize the library m-services to that particular user. The second tier consists of the InfoStations themselves, deployed throughout the campus. These wireless nodes provide users with mobile access to the library m-services through wireless connections. The InfoStations also house cached profile repositories, which contain information relating to all registered users. The InfoStation-based system is organized in such a way that if an InfoStation cannot fully satisfy the user service request, the request is forwarded to the InfoStation Centre situated at the core of the architecture. The InfoStation Centre houses a repository of up-to-date copies of all master profiles relating to users and services. Figure 6.1 illustrates some of the main components within each level of the architecture. The InfoStation Centre's design is based on suggestions described in Andersson et al. (2006).

Library-based services

A multitude of institutions around the world have harnessed the capabilities of technologies such as SMS/MMS and podcasting to supplement existing teaching and learning schemes. SMS has proved to be most useful when used in conjunction with the traditional teaching practices. SMS/MMS is a technology that has been embraced by mobile phone users, especially within tertiary education institutions, where virtually every student has access to at least one mobile device. In harnessing this ubiquitous technology, a resource that has been looked upon as a nuisance within educational institutions becomes a tool to better achieve educational goals. This is especially the case when interacting with large groups of students and in disseminating information. SMS/MMS technology has been used in a number of innovative and creative ways so as to enhance the effectiveness of large classes and to enrich the learning experience of students (Kinsella, 2008). Within large classes, it is often difficult for educators to judge the assimilation of information by learners. The system proposed by Kinsella enables students to send anonymous text messages to a particular phone number. These messages are then displayed on the screen behind the lecturer. This enables students to interact with each other and with the lecturer, and allows the lecturer to respond to student observations, questions and comments in a controlled manner. This example illustrates the use of SMS technology to convey information from students to the lecturer. Moreover, this technology can aid the flow of information to large groups, especially in the conveyance of admin-istrative information (e.g. events taking place on campus, notifications of project submission deadlines, etc.).

Podcasting is another technology that has proved useful in supple-menting the traditional learning experience. Podcasting describes the provision of sound files (e.g. MP3) that can be downloaded by users and played on mobile devices. Podcasts are used regularly within mainstream media (e.g. radio shows). In the educational domain, increasing numbers of educators are seeing podcasts as a means of enhancing the effectiveness of traditional learning practices and supporting distance education. Lecturers can take audio recordings of lectures and make them available for students to review and study at their own pace. An example of such use of podcasting is outlined on the Dukecast website[1] and in Duke (2005), describing how Duke University provided students with iPods equipped with voice recorders. By being enabled to make their own recordings of lectures, students can benefit from being able to structure their learning to a pace that suits them as individuals. While these technologies have been used in catering to a great many educational circumstances, there are also a multitude of possible applications within the library domain. The aforementioned architecture is utilized in order to evolve the existing library services, providing and facilitating a mobile component to these services.

Library catalogue, loans and reservations

This service builds on the existing database and cataloguing system, allowing students to access the library catalogue while on the move. A benefit of this service would be a reduction in queues to access library computers to search for resources. In addition, users can access up-to-date information as to the availability of resources. The service also allows users to monitor the status of resources that they have on loan from the library and to monitor when a resource is due to be returned and, indeed, the fines that may be incurred should a return date be missed. In future, the extension of this service with the Business Support Domain will allow users to pay any library fines they may incur directly from their mobile devices. This service will also enable users to reserve library resources currently on loan to other users, and to receive automatic notification when the resource becomes available.

Interactive library map

In collaboration with the library catalogue service, this service facilitates the quick and efficient location of resources within the library. Users are provided with specific directions to the collections of materials that are most suitable to them (e.g. Science, Engineering, Languages), or indeed to the location of a very specific resource that the user may have requested through the library catalogue. In the delivery of this service, the content is adapted and customized according to the capabilities of the user device and user preferences. If the user is using a device with limited graphical display capabilities, the service will be provided in a simple format which 'best' suits the device (i.e. textual format). However, from a device with a much higher range of capabilities (e.g. a laptop), the complete hypermedia format of the service may be accessed. This will encompass a interactive graphical representation of the library building, detailing directions to the specific resource. This adaptation and customization of the services addresses the need for the system to provide information to facilitate a wide variety of devices with varying capabilities, while still delivering the service in the 'best' possible manner. In conjunction with the library catalogue service, this will allow users to make more efficient use of their time in the library. It will also reduce the workload of library staff, as even first-time library users will be able to navigate the library with ease.

Service provision

Figure 6.2 illustrates a sample interaction between the entities involved in the provision of library-based services. When users enter within the range of a library InfoStation they go through an Authorization, Authentication and Accounting (AAA) procedure. The PA also conveys information detailing the user device capabilities and updates of the user profile and user service profile. If the InfoStation has no record of this user profile in its cached repository, it forwards the request on to the InfoStation Centre. The InfoStation Centre, if necessary, creates a new account for that user or, if it has records of that user, completes the AAA transaction and sends an acknowledgement back to the PA. On completion of the AAA procedure, the InfoStation analyses the user profile and device capabilities and draws together a list of recommended resources that are applicable to the user. Details of the location of these resources as well as interactive directions to their location are provided to the user's PA. The PA maintains the details and location information of recommended resources in its cache, in case the user requests them. If the user chooses to access the library catalogue service, s/he may specify a number of search criteria, such as keyword, title, author, year of publication, or indeed which collection to search (Main Catalogue, Journals, Special Collections, etc). Once entered, these search criteria are used by the InfoStation to sift through the library

Figure 6.2
Interactions in library-based
services

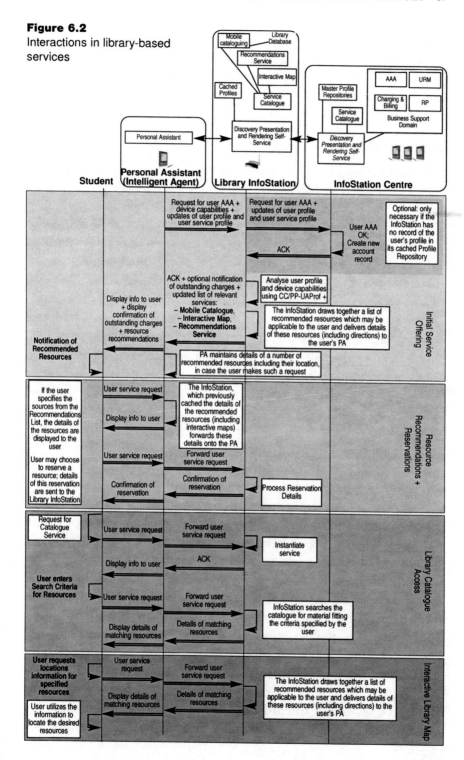

databases and find material that 'best' matches the user's search. Once the user has chosen the resource s/he wishes to obtain, the InfoStation provides details of the resource's location within the library building, again drawing on the Library Interactive Map service to provide the user with directions.

Contextualization of services

With the increasing level of diversity of end-user devices, the service content must be translated to meet the requirements of increasingly heterogeneous interfaces. In order to handle this diversity, the PAs must have an explicit knowledge of capabilities of the devices on which they are based, and of course the ability to convey this information to the service provider. In order to provide a structure through which the intelligent agent can efficiently employ and convey essential Capability and Preference Information (CPI), we utilize the Composite Capability/Preference Profile (CC/PP) (Gil et al., 2003; Tran et al., 2003; W3C 2006) standard, and in particular its User Agent Profile (UAProf) (OMA, 2006) module. The CC/PP specification defines a structured framework through which devices can define critical criteria relating to their capabilities. The UAProf builds on this, providing a concrete implementation of the CC/PP, facilitating the end-to-end flow of CPI throughout the system. Context definition is the first step in the contextualized and personalized delivery of services.

Utilizing the CPI contained within user profiles, the service provider (i.e. InfoStation) can tailor its service offer to the requested delivery context. The UAProf represents CPI in a two-level hierarchy consisting of various *components* (defining different characteristics of the device) and associated *attributes*. The details specified within an instance of the CC/PP-UAProf and its components enables an InfoStation to dynamically adapt and customize m-services according to the specifications of that device. One of the most important components taken under consideration is the *hardware* component of the device. This details the device's hardware characteristics, providing information about the screen size, imaging capabilities, processing power, input/output methods, etc. An InfoStation uses this information to efficiently adapt particular service content to the hardware environment on that device. An example of how a service may be adapted according to hardware criteria would be, for example, to take into account the *ScreenSize* attribute. This attribute places constraints on the amount of information that can be shown on the screen of the device. Figure 6.3 illustrates a sample of a device's hardware component with attributes which may each have a bearing on the appearance of the final delivered contextualized service content.

```
<prf:component>
 <rdf:Description rdf:ID="HardwarePlatform">
 <rdf:type rdf:resource="http://www.openmobilealliance.org/tech/profiles/UAPROF/
ccppschema-20021212#HardwarePlatform" />

 <prf:ScreenSize?176x220</prf:ScreenSize>
 <prf:Model?V3i<prf:Model>
 <prf:ScreenSizeChar?17x11</prf:ScreenSizeChar>
 <prf:BitsPerPixel>16</prf:BitsPerPixel>
 <prf:ColorCapable?Yes</prf:ColorCapable>
 <prf:TextInputCapable>Yes</prf:TextInputCapable>
 <prf:ImageCapable>Yes</prf:ImageCapable>
 <prf:Keyboard>PhoneKeypad</prf:Keyboard>
 <prf:NumberOfSoftKeys>2</prf:NumberOfSoftKeys>
 <prf:PointingResolution>None</prf:PointingResolution>
 <prf:CPU>Motorola LTE-ARM7TDMI-S</prf:CPU>
 <prf:Vendor>Motorola</prf:Vendor>
 <prf:PixelAspectRatio>1x1</prf:PixelAspectRatio>
 <prf:SoundOutputCapable>Yes</prf:SoundOutputCapable>
 <prf:StandardFontProportional>No</prf:StandardFontProportional>
 <prf:VoiceInputCapable>Yes</prf:VoiceInputCapable>
 <rdf:Description>
 <prf:Component>
```

Figure 6.3 CC/PP hardware component

Other important components that have a bearing on service delivery are the software component (Java platform, etc.), network component (supported bearer services, Bluetooth version, etc.), and WAP component (WAP version) – all enabling the InfoStation to better adapt a service to whatever environment it may be called upon to operate in and, as such, ensure the optimal conditions for the service to achieve its stated goals. Each of these components serves to ensure that the services can be offered across a multitude of platforms and user environments, providing the system with a level of device independence. Of course there is no restriction to using just these static, pre-defined components. We ourselves can define any components that may be deemed necessary for the successful implementation of contextualized and personalized services. Within this system we must define a number of user-related attributes such as user name, programme, department, etc. This innovation in the use of the UAProf and the creation of user-based attributes has a major bearing on the services offered to an individual, and indeed on the personalization of the services that are ultimately advertised to the user. Figure 6.4 is an example of how a component and group of attributes relating to a particular user may specify vital information about that individual.

```
<prf:component>
<rdf:DescriptionrdfID="UserPlatform">
 <rdf:typerdf:resource="http://www.ece.ul/ie/trc/profiles/UAPROF/ccppschema-1#
UserPlatform"/>
<prf:Name>JohnDoe</prf:Name>
<prf:StudentID>0123456</prf:StudentID>
<prf:Faculty?ECE</prf:Faculty>
<prf:Course>Electronic Engineering</prf:Course>
<prf:Year>4</prf:Year>
<prf:Classes>
    <rdf:Bag>
            <rdf:li>CE4517</rdf:li>
            <rdf:li>CE4607</rdf:li>
            <rdf:li>CE4717</rdf:li>
            <rdf:li>CE4817</rdf:li>
            <rdf:li>CE4907</rdf:li>
            <rdf:li>CE4607</rdf:li>
    <rdf:Bag>
<prf:Classes>
<prf:Advisor>Dr. Ivan Ganchev</prf:Advisor>
<prf:email>0123456@STUDENT.ul.ie</prf.email>
<prf:QCA>3.47</prf:QCA>
<prf:FYP>JD09</prf:FYP>
<prf:FYPSupervisor>Dr. Ivan Ganchev<?prf:FYPSupervisor>
<prf:FYPTitle>Design and Implementation of an Animated Interactive
Tutorial</prf:FYPTitle>
</rdf:Description>

</prf:component>
```

Figure 6.4 CC/PP user-specific component

In the given sample component, an attribute such as 'Faculty' can have a major bearing on the type of services being offered to a particular user. For example, the services used by members of the Business Faculty will generally differ greatly from those being used by members of the Science and Engineering Faculty. Indeed, by allowing for the specification of the various classes being taken by the user, services can be directed at even more specific ranges of users. It is essential for these factors to be taken into account, in order to avoid InfoStations unnecessarily advertising irrelevant service and content to users. Essentially, this innovative approach to user preference dissemination facilitates the successful delivery of highly-personalized services to the client device.

Conclusions

In the course of this chapter various aspects regarding the evolution of library services, based on an underlying InfoStation-based architecture have been discussed. The chapter has described the main elements of a service architecture needed to support the expansion of existing library-based services into the mobile domain, based on a model proposed as part of the Distributed e-Learning Centre (DeLC) initiative. The enhancements facilitated to traditional library services and educational practices through the incorporation of SMS/MMS and podcasting technologies have been explored, highlighting how these technologies have previously been adopted into the spheres of learning, and the advantages that these technologies can afford to library users and to learners in general. Two supported library services, the 'Interactive Library Map' and 'Library Catalogue, Loans and Reservations' have also received particular attention, as have the amalgamation of these services in order to facilitate an automated recommendations service. Also discussed were the ideals pertaining to the provision of increasingly contextualized services, and the ability to afford system users a great deal more ownership and control over their interactions with the presented services. Trends towards student 'ownership' of their learning interactions, and how the capabilities of these mobile processes can ensure better dissemination and reorganization of the large volumes of administrative information that tertiary education institutions deal with on a daily basis, have been dealt with in detail.

Acknowledgements

This publication has been supported by the Irish Research Council for Science, Engineering and Technology (IRCSET) and the Bulgarian Science Fund (Research Project Ref. No. dO02-149/2008).

Note

1   Dukecast,
    http://dukecast.oit.duke.edu/.

References

Andersson, F. et al. (2006) Mobile Media and Applications – from concept to cash, Wiley.

Duke (2005) Duke University iPod First-Year Experience Final Evaluation Report, http://cit.duke.edu/pdf/ipod_initiative_04_05.pdf.

Ganchev, M. et al. (2008) Integrating the Educational Support Architecture in an E-Services Paradigm: the m-learning approach. In: Needham, G. and Ally, M., *M-libraries: libraries on the move to provide virtual access*, Facet Publishing.

Ganchev, O'Droma et al. (2008a) InfoStation-Based Adaptable Provision of m-Learning Services: main scenarios, *International Journal 'Information Technologies and Knowledge'*, (IJ ITK), 2 (5), 475–82.

Ganchev, O'Droma et al. (2008b) On Development of InfoStation-based mLearning System Architectures. In 8th IEEE International Conference on Advanced Learning Technologies (IEEE ICALT08), Santander, Cantabria, Spain.

Ganchev, O'Droma et al. (2008c) InfoStation-Based Library Information System. In 8th IEEE International Conference on Advanced Learning Technologies (IEEE ICALT-08), Santander, Cantabria, Spain.

Gil, G. et al. (2003) Delivery Context Negotiated by Mobile Agents Using CC/PP, *Lecture notes in computer science*, 99–110.

Kinsella, (2008) Many to One: using the mobile phone to interact with large classes, *British Journal of Educational Technology*.

OMA (2006) User Agent Profile V2.0, Open Mobile Alliance.

Tran, B. et al. (2003) Composite Capability/Preference Profiles (CC/PP) Processing Specification, Sun Microsystems.

W3C (2006) Composite Capability/Preference Profiles (CC/PP): Structure and Vocabularies 2.0, World Wide Web Consortium (W3C)

7

Bibliographic ontology and e-books

Jim Hahn

Abstract

This chapter consists of a theory paper on the conceptual model featured in the Functional Requirements for Bibliographic Records (FRBR) (IFLA, 1998) and its use in providing a framework for acquiring content used on mobile devices. FRBR is promulgated for the convenience of users of bibliographic systems. The paper is a proof of the concept that the IFLA bibliographic ontology remains a powerful tool for the conceptual understanding of a type of mobile digital content: e-books. The *manifestation* entity is of considerable significance for collecting e-content. (Previous work in this area includes theorization of items in a digital world (Manoff, 2006; Floyd and Renear, 2007).) The paper will present the idea of using this *manifestation* entity and the relationships it holds with other *manifestation* entities as a way to conceptualize digital content.

These issues can be understood in practice by drawing on the work of Tillett (1991), which enquired into the bibliographic relationships of works in cataloguing practice. In a study surveying multiple past cataloguing codes, Tillett (1991) found that cataloguing rules are cognizant of equivalent manifestations and make use of such bibliographic relationships in cataloguing practice. The significance of this paper is to present the FRBR conceptual model as a guide to collection developers.

Introduction to FRBR

The Functional Requirements for Bibliographic Records feature a conceptual model for helping librarians to articulate 'what exists' in the bibliographic world. This ontology is high level and terminological, rather than a more specific, assertional-level ontology. It defines the entities, attributes and relationships of the bibliographic universe. This definition is independent of any implementation or bibliographic system. One conceptual articulation of the FRBR ontology is made through an entity relation diagram (IFLA, 1998, 14). The entities of the bibliographic universe set forth are certainly understandable to the general library science practitioner. These definitions make sense in terms of thinking about what our users are asking for when they ask us to order forms of content that would seem to be novel to library and information science.

The focus of this chapter is the Group 1 entity sets, articulated in the *Functional Requirements for Bibliographic Records: final report* as serving to 'represent the different aspects of user interests in the products of intellectual or artistic endeavour' (IFLA, 1998). The Group 1 FRBR entities include these concepts:

Work: 'a distinct intellectual or artistic creation'
Expression: 'the intellectual or artistic realization of a work'
Manifestation: 'the physical embodiment of an expression of a work'
Item: 'a single exemplar of a manifestation'.

The *work* entity is an abstract idea, it is the artistic or intellectual content of the book; the *work* entity is somewhat difficult to envision – the concept of the work is essentially the artistic idea realized when the book is read. The *work* entity is not something you can point to as a physical thing. The *expression* refers to the written content. Common reference to different translations of the *work* is one example of an everyday use of the *expression* term. The *manifestation* entity can be thought of as the particular edition of a book. Manifestation entities also include formats of the expression. *Manifestation* is identifying a 'physical embodiment' (IFLA, 1998, 21). *Items* progress in a similar fashion to the manifestation entity, in that we are describing physicality. The item entity set is different from manifestation in that a specific copy is sought. An example of a library user seeking an *item* is a case of the book containing the unique library barcode. It is a barcode that only this book has. This unique barcode is different from the ISBN number of a book, which references a specific *manifestation*. It is hoped that this model will make the bibliographic universe clearer; with these entities and the relationships between them, we can better articulate what is meant by the term 'book', that is to say, are we discussing the intellectual idea (work), a specific translation (expression), an edition of that translation (manifestation) or a unique object (item). The FRBR conceptual model is a step toward providing clarity for the vocabulary we use to describe the bibliographic world (Tillett, 2003, 1, 11).

Bibliographic relationships in practice

The FRBR is articulated as a conceptual model independent of any specific bibliographic system or a specific implementation. As a practice-oriented profession, it makes sense for us to look at past cataloguing practice to see the historical continuities among bibliographic relationships and the FRBR conceptual model. Tillett (1991) provides a review of what relationships exist in actual cataloguing practice in the article 'A Summary Treatment of Bibliographic Relationships in Cataloguing Rules'. Not only is Tillett (1991) concerned with a review of past cataloguing codes employed for describing bibliographic relationships; there seems to be a concern for conceptual models for bibliographic systems and what those models might articulate: 'One part of that model would be the various relationships we want to express, including bibliographic relationships, access point relationships, etc.' (Tillett, 1991, 393). The taxonomy of bibliographic relationships that Tillett uncovered in this study includes:

1 Equivalence relationships
2 Derivative relationships
3 Descriptive relationships
4 Whole–part (or part–whole) relationships
5 Accompanying relationships
6 Sequential relationships
7 Shared characteristic relationships (Tillett, 1991, 394).

The equivalence relationships are useful for our enquiry into how e-books are placed within the bibliographic model. In Tillett's study, equivalent relationships are described as 'those that hold between exact copies of the same manifestation of a work, or between an original work and reproductions of it, as long as intellectual content and authorship are preserved' (Tillett, 1991, 394). For the purposes of mobile content, the manifestation (or edition) of an e-book holds an equivalent relationship to the print manifestation. The FRBR *Final Report* is cognizant of the need to model manifestation to manifestation relationships in bibliographic systems (IFLA, 1998, 61–2).

Where are we?

This chapter is exceedingly abstract, especially if the Functional Requirements for Bibliographic Records are new to you. To restate what has been covered and where we are going: thus far we have introduced the FRBR Group 1 entities, and understand these entities to define a conceptual model of the bibliographic universe that helps us to understand entities, attributes and their relationships in bibliographic systems. We have also shown that there exists a history of these bibliographic relationships in cataloguing practice. We want to show now that the manifestation entity holds equivalence relationships with the manifestation entity describing e-books. To make this final connection we now return to the FRBR *Final Report*.

Manifestation relationships

FRBR articulates the boundaries of manifestation entities – which advise that a new manifestation of a work exists when the physical medium changes; so the e-book is not actually the same manifestation as the print resource, but holds a relationship to the printed manifestation of a work (IFLA, 1998, 22). It follows from the manifestation relationship that, as a librarian enquires into what is being purchased when an e-book is ordered, she will see that the print book and the e-book are manifestations of the same work. The collection decision will be made accordingly.

The Web Ontology Language[1] includes tools for making equivalent relationships using elements of RDF (Resource Description Framework)[2] and RDF-S (RDF Schema)[3] and first order logic. These are tools that librarians may use to describe the relationship between bibliographic content and other content accessed through the world wide web. This mark-up would be used for computationally processed 'meanings' of entities and their relationships. These mark-up tools are not the focus of this chapter, but are useful for further study in ontology, especially to the librarian seeking to document an assertional-level ontology. Useful introductory surveys include the world wide web consortium documents and specifications found at www.w3.org.

Equivalent relationships among manifestations are advocated here as being a means to describe how the content accessed for mobile devices may fit into an already existing LIS paradigm. The e-book is not such a unique resource that it exists completely outside of the domain of the relationships we want to model in our bibliographic systems. Think of other formats, like microfilm, DVD, VHS; these formats use electronic devices to display content, but librarians are still able to describe attributes and situate these within the bibliographic sphere. E-books are a format, often available to our users as licensed content.

Conceptual models can be used to model an information system (like a database, or online catalogue); but they can also be used to articulate what exists and the thesis here: to provide a way to understand what the library seeks to acquire. As the domain of mobile digital library content grows, there will be a need to show what it is that defines library content. In specifying the models over what content the library would purchase, a conceptual model like FRBR can be utilized to articulate what content the library currently provides access to, and in what form.

If we return to an earlier thread of this chapter, modelling entities of the bibliographic world are helpful in providing clarity for the users of bibliographic systems. If the conceptual model helps to bring clarification over the range of content the library seeks to acquire and provides access to, then it should be referred to by those who develop collections.

Conclusion: the significance of FRBR

The licensed content of information the library provides is vastly fragmented. This access does exist across many access platforms, databases and vendors. It may be the case that computationally derived searching and indexes like Google have made a formidable case against the ordered conceptual models of the LIS world. Further fragmentation is evident in the ways in which only pieces of content may exist in the library collection.

The FRBR includes whole–part relationships which can accommodate the description of chapters of e-books. Another practical application of whole–part relationships is within the domain of music content. Apple's iTunes store, in particular, has made the acquisition of pieces of music albums a popular purchase. We may have only a part of an e-book, or we may wish to describe the parts of an e-book in bibliographic systems. Chapter 5 of the final report (IFLA, 1998, 55) describes the various bibliographic relationships of the FRBR model.

Acknowledgements

This chapter would not have been possible without the supportive environment of the Ontology Development course LIS-590OD, Spring 2008, at the University of Illinois at Urbana-Champaign, led by Professor Allen Renear, Graduate School of Library and Information Science. Many thanks also to the conference attendees at M-Libraries 2009 for helpful feedback on a presentation on this topic.

Notes

1   www.w3.org/TR/owl-features/.
2   RDF can be described as a data model (Antoniou and van Hamelen, 2008, 66).
3   RDF-S actually describes the vocabulary of an RDF data model (Antoniou and van Hamelen, 2008).

References

Antoniou, G. and Van Hamelen, F. (2008) *A Semantic Web Primer*, Cooperative Information Systems, MIT Press.

Floyd, I. and Renear, A. (2007) What Exactly is an Item in the Digital World? *Proceedings of the American Society for Information Science and Technology*, 44 (1), 1–7.

IFLA (International Federation of Library Associations and Institutions) (1998) *Functional Requirements for Bibliographic Records: final report. Recommended by the IFLA Study Group on the Functional Requirements for Bibliographic Records*, www.ifla.org/files/cataloguing/frbr/frbr_2008.pdf.

Manoff, M. (2006) The Materiality of Digital Collections: theoretical and historical perspectives, *portal: Libraries and the Academy*, 6 (3), 311–25.

Tillett, B. (1991) A Summary of the Treatment of Bibliographic Relationships in Cataloguing Rules, *Library Resources & Technical Services*, 35 (4), 393–405.

Tillett, B. (2003) What is FRBR (Functional Requirements for Bibliographic Records)? *Technicalities* 23 (5), 1 and 11–13.

Suggested reading

Kent, W. (2000) *Data and Reality*, 1st Books Library.

Renear, A. and Salo, D. (2003) Electronic Books and the Open eBook Publication Structure. In: Kasdorf, E. (ed.), *The Columbia Guide to Digital Publishing*, Columbia University Press, 455–520.

8

QR codes and their applications for libraries – a case study from the University of Bath Library

Kate Robinson

Figure 8.1
An example of a QR code

A QR (Quick Response) code is a two dimensional bar code that can be read on a device such as a mobile camera phone, a laptop or other computer. When read, it allows the user to undertake an action such as reading text, accessing a website or texting a number. In other words, a QR code links the physical world (poster, printout, room, physical object) to the electronic (web resource) and facilitates communication (SMS message, phone call), adding significant value by improving accessibility to information for those using mobile devices. The University of Bath Library has been exploring and testing library applications for QR codes, with the emphasis very much on experimentation and discovery. An example is given in Figure 8.1 above.

As a technology, the generation and reading of QR codes on a mobile device is becoming straightforward, with web-based generators becoming more available and easier to use, and the appropriate software pre-installed on many camera phones. This technology is well established in countries such as Japan and is starting to enter the mainstream in the UK across a number of sectors, including media and marketing. QR codes are beginning to appear on flight and train tickets and to provide links to advertising and promotional material through posters, business cards and even coffee cups. At Bath we have found that they are still very much an emerging technology, with relatively low student awareness.

QR codes can be read from a printed or an electronic image. To read a QR code you need QR reader software installed on a compatible mobile camera phone or computer, so that when you scan or photograph the code the software can decode it. This decoding facilitates an action such as opening a browser at a web page, opening a text messaging program pre-populated with a number to text, or providing some other textual information. So, for example, you can link a physical instance of the code, e.g. from a poster, with a web resource to provide additional information.

QR codes join things up by removing barriers, which is very familiar territory for librarians, who continually move things around behind the scenes so that users can find what they need. There are clearly many potential applications for these codes and, as their production is relatively inexpensive, they are also appropriate for temporary use. At Bath, our main focus has been on the library catalogue. We have attached QR codes to library catalogue records, thereby allowing students to capture biblio-graphic and location information on their mobiles.

The QR codes are created dynamically using the Google Chart API (Application Programming Interface) to generate the QR code and JavaScript to include it in the page. It should work with any database-driven website. The QR code encodes the author, title, location and call number of the resource as a text string, enabling a user to display these details on their camera phone before going to locate an item on the library shelves. The QR code is displayed only if a copy is available on the open shelves or, failing that, in the short loan collection. This is a clear example of linking the virtual to the physical. Instead of searching the catalogue and writing information on paper, scanning the code directly from the screen to a phone allows a user to locate the resource and to save the bibliographic and location details for later use.

QR code use does not generate statistics, so our initial evaluation of their use has been mainly anecdotal. We found that a number of students recognized QR codes and many of them liked the concept of being able to scan something to complete a task. Previous research by our e-learning colleagues had suggested that many students would be able to use a QR reader on their phone. It soon became evident that it can be difficult to install a reader and that some simply do not work effectively for this purpose. We found that a much smaller number of people than we had originally envisaged were able to scan QR codes using their phones.

With this in mind, I see support as one of the key elements of any future

QR initiative at Bath. As well as offering hands-on support to install the appropriate readers, we should enable and provide opportunities for students to help each other. It would also be sensible for information regarding QR code reading and readers to be integrated into the university's existing IT help infrastructure, along with mechanisms for the university to bear the cost of any resulting data and file transfer charges. The other area to address is one of motivation. Until QR readers become ubiquitous there needs to be a good reason for users to take the trouble to load and use them. Locating books and building lists of references alone does not really add sufficient value to persuade library users to engage with this. A consistent use of QR codes across campus may well be more persuasive in future.

We have seen that there is a growing expectation that services can be delivered to mobiles. It makes sense for us to experiment with technologies such as QR codes to join up library services with the technology and equipment students use. This is just another tool which may work for some. It is still early days for QR use in the UK and neither the technology nor its use are mainstream as yet, but we now know that we can generate QR codes and we have identified some uses for them. Our intention is to continue to experiment in order to see whether and how they begin to add value for our users.

9

A tale of two institutions: collaborative approach to support and develop mobile library services and resources

Hassan Sheikh and Tony Tin

Introduction

The Open University (OU), UK and Athabasca University (AU), Canada are both world leading distance-learning institutions. The OU currently has more than 200,000 students studying various undergraduate and postgraduate courses, while AU serves over 37,615 students and offers over 700 online courses. For distance learning, it is very important for online resources to be made accessible to as wide a range of users and devices as possible. It is essential that both existing and emerging technologies be implemented for effective deployment, delivery and support to remote students. In order to keep up to speed with mobile learners' needs, several mobile learning initiatives have taken place within AU and OU. Both AU and OU libraries have initiated a strategic partnership to share expertise, knowledge and development work about mobile library services and resources.

This chapter will focus mainly on the mobile services development work at the Open University, UK but will also highlight the collaborative work being undertaken between Athabasca University and the Open University and the benefits, opportunities and challenges in developing mobile library services and resources.

Mobile learning and mobile library services

Mobile learning has for some time now had a high profile within the OU, as evidenced by the work of colleagues in our Institute of Educational

Technology, and because mobile access to our Virtual Learning Environment (VLE) was planned from the beginning of the project to implement Moodle (the open-source VLE used by the OU). As we are a distance learning institution, mobile access to course materials seems like a natural extension of our online course offerings. Seventy percent of OU students are in full-time employment. Many of them have families and other responsibilities as well. Giving students the opportunity to access their course materials or library resources anytime, anywhere may assist them in scheduling study time into their busy lives. We also offer a range of work-based learning opportunities, where students can benefit from being able to review learning materials or reflect on practice at work.

Why mobile services are important for distance learners

Mobile technologies have provided unique opportunities for educators to deliver educational materials efficiently and to support the cognitive and social process of students' learning. Educational materials can be delivered to students through mobile devices. Students can communicate and interact with their peers and educators in real time, using mobile technology. Mobile technology can also be integrated into curriculum design to improve interactivity in the classroom. Applications of mobile technology in education can provide benefits to both students and educators. Mobile technology provides greater flexibility in student learning.

Mobile phones and other portable devices are personal tools that elicit intensely personal reactions. Mobile learning challenges us to look beyond the apparent limitations of small devices and the annoyance they can sometimes cause, to understand their positive role in connecting people, taking advantage of location and making learning more accessible. Many OU students combine work and study; consequently, learning in a number of places, or on the move, becomes a habit. Informal learning and social interactions are also increasingly recognized as important components of a person's 'learning life'. Academic and support staff are part of this revolution (Pettit and Kukulska-Hulme, 2007).

The latest trends in the mobile phones market

The touch-screen phone market (Apple's iPhone and Google's Android) is expanding quite rapidly, especially in the US and Europe. According to the latest forecast by BMO Capital's Keith Bachmen (Hadar, 2008), Apple will sell 26 million iPhones (4.3 million more handsets than the forecast for 2009) by the end of 2010. As iPhone and Android handsets are becoming relatively affordable, we (at the OU Library) have been observing (through our Google Analytics[1] website stats) a steady growth in the number of our students using them and accessing the electronic resources through these leading-edge mobile phones. Based on the growing number of our users, there is a stronger need to develop a specialized version of the applications (and websites) for the latest class of touch-screen phones so as to deliver a better-quality interface for our e-resources and improve the user experience, by making use of the larger screen and the intuitive interface of these touch-screen devices.

What services users want on their mobile phones

At the OU we scan continuously in order to decide which services to develop. We have also undertaken some research into user requirements, such as the *M-Libraries: information on the move* (Mills, 2009) research undertaken as part of the Arcadia Programme, and observation of users engaging with mobile information management skills learning objects.

The key recommendations from this user requirements research are:

- to ensure that the library website continues to be able to resize to fit smaller screens and that the opening hours and contact details are easy to find from the home page
- to pilot the prototype mobile OPAC interface being developed by the OU Library's systems development team and share the development with other libraries that are also using Voyager from Ex-Libris
- to pilot text alerts for loaned items that are due for renewal or overdue, and for requested items that are available for collection
- to pilot text reference services
- to consider adapting Athabasca University's Digital Reading Room to allow mobile access to our electronic resources and to ensure it works on multiple platforms, including Apple's iPhone and Google's Android
- to develop a project proposal to work with SCONUL (Society of College, National and University Libraries) and Oxford University

Computing Services on providing mobile access to the SCONUL database of libraries; this location-aware service would enable students to automatically get a list of libraries near them, depending on their current location

■ to pilot library audio tours for new library users and visitors by making them available for download through the library website and pre-loaded on MP3 players for loan.

OU students use a wide range of mobile phones and hand-held devices to access the library website. These mobile phones vary from lower-end simple mobiles such as Nokia 1112 or Samsung E250 to leading-edge smart and more advanced phones like Apple's iPhone or Google's Android. It is therefore important for us to develop the key functions of the website to work effectively and accurately on phones with limited capabilities as well as on the latest high-end phones. Planned improvements to the website also include providing a link enabling visitors to switch between the mobile version and the full version of the site.

Open University UK's current work on the development of mobile library services

Partnership with Athabasca University Library

Following a visit to the OU by Mohamed Ally of Athabasca University in 2007, a partnership was formed between the two university libraries on developing library services and content delivery systems for mobile and hand-held devices. This partnership has provided an opportunity for both parties to work together strategically and to develop some innovative mobile library services. One of the significant outputs of this partnership has been the successful organization of both the First and Second International M-Libraries Conferences, which have been jointly organized by the Open University, UK and Athabasca University, Canada with other educational institutions.

The OU Library uses the ADR (auto detect and reformat) software which was developed by the AU Library development team. The ADR software allows developers to make websites and other online content suitable for viewing on a small screen. The software automatically detects the type of mobile device used to connect to the website and both renders and optimizes the contents to fit appropriately on the mobile screen, by changing the layout template and style-sheet. The advantage of this approach is the ability to use the same content to be rendered on two different display models (one is for normal screens and the other is for smaller size screens) and it has saved our content authors from writing two separate versions of the same content.

The mobile version of the OU Library website also offers a customized home page and simpler search interface for users. There were more than 21,500 page visits through the mobile interface between its launch in October 2007 and July 2009. The mobile interface supports most standard mobile devices such as PDAs, palmtops, Blackberries, HP iPAQs and Nokia N95s.

The same ADR software has enabled us to develop a mobile Learning Object Generator (LOG), which has been used to create short revision modules for the OU Library's online information literacy tutorial, Safari and the Information and Knowledge at Work resources (iKnow).[2] The LOG system enables instructional designers to author mobile-friendly learning objects and package them into a downloadable zip file. The package zipped file conforms to IMS content packaging standard [4][3] and it contains the XML manifest file with all the content resources bundled together.

Tracking user behaviour through Google Analytics

We have been using Google Analytics since 2007 to track the OU Library website traffic (for both desktop and mobile versions) and to analyse the behaviour of users visiting the OU Library website. The Google Analytics tool provides a lower barrier to entry, in terms of allowing inexperienced users to easily create eye-catching graphical reports that provide information on the source of traffic (where visitors came from), what pages and areas they visited, how long they stayed on each page, how deep into the site they navigated, where their visits ended, where they went from each page and so on.

Google Analytics statistics have revealed that the most popular areas for the library website's mobile users are the home page, Contact Us, Opening Hours, News, Jobs and Events at the OU Library and Search the Library collections. These findings have also been confirmed during the Arcadia research project, when users were asked about the top library services they would like to access from their mobile phones. Based on these findings, we are currently revamping the mobile OU Library website and will focus on improving these key areas. Figure 9.1 shows how the mobile version of the OU Library website home page has evolved as a result of the findings discussed above.

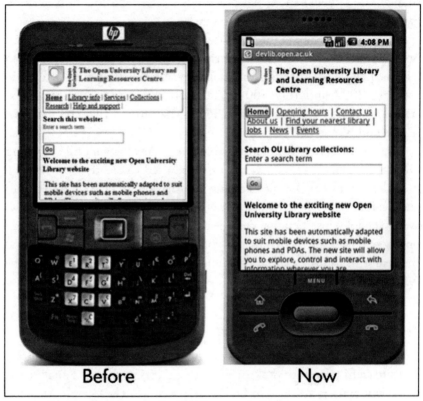

Figure 9.1 Development of the OU Library website home page

Survey of OU students' requirements

The OU student respondents to the *M-Libraries: information on the move* survey indicated that the main areas of the website to which they would be likely to want access through their mobile phones would be the opening hours (75.9%) and the library's contact information (70.6%), so the mobile interface will provide links to those pages from the website's home page. Students at Cambridge University indicated that, in addition to these areas, they would want quick access to the location details of the library.

Usability testing

The OU Library plans to undertake usability testing of the website, including the mobile version, in the last quarter of 2009. The results of this testing will help us to further evaluate and refine the mobile-friendly version of the site. The mobile Learning Objects are being evaluated through observation of users, and a programme of work-based testing in the case of iKnow. They will also be added to our Google Analytics tracking in the future.

The OU Library's strategy includes developing user-centred resources and services and supporting the delivery of e-learning. Our mobile services and skills offerings will provide OU students with a greater degree of flexibility in their study and enable them to access our content whenever and wherever they choose.

Implementation of m-library services based on the gathering of user requirements

It has been over two years since we started developing mobile services at the OU Library and we have achieved a number of milestones so far, including:

Adapting auto detect and reformat: ADR renders and optimizes the website content to fit nicely on smaller screens and includes most makes and models of the phone in its detection engine. Since the AU/OU collaboration was initiated in 2007, we have been actively working on enhancing and adding new functionality to the ADR software, such as screen size detection, stripping out the images from content, validating html tags, and the ability to render new style-sheets. This is a cost-effective approach in comparison to hand-crafting a separate mobile website. The m-libraries survey results show that a low percentage of users currently access the mobile internet.

Piloting a consolidated search function for mobiles: We have prototyped a cut-down version of the search function to display results with simpler metadata fields, such as the result title (and a link to the title) and result score. The initial mobile version (developed in 2007) of the search function included only a single collection, i.e. the OU Library website. However, the current version we are working on is a consolidated search solution that will include nearly all our e-collections (Library catalogue, e-Journals, Open Research Online repository and third-party licensed databases), to be searchable through a single search box. Figures 9.2 and 9.3 illustrate the technical architecture and the front end of the consolidated search solution.

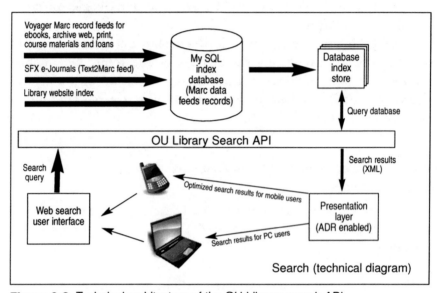

Figure 9.2 Technical architecture of the OU Library search API

Although the search interface is able to show the search results in a mobile-friendly manner one of the challenges is that the majority of third-party content is not yet optimized for mobiles, so if users click on any of the search result links they may not be able to read the content of the target pages. We have been in discussion with some database vendors about the possibility of optimizing their content for mobiles, but for the majority of them apparently either they don't have capacity to do so or it isn't a top priority. Another option is to redirect users to view those target pages through Skweezer,[4] although it doesn't guarantee that the website will render correctly on the mobile phone.

It is challenging to render lengthy content on smaller screens. It has been observed in our usability evaluations that users don't like to read large chunks of text on smaller screens. We have been working with our content authors to help them with writing concisely and to the point.

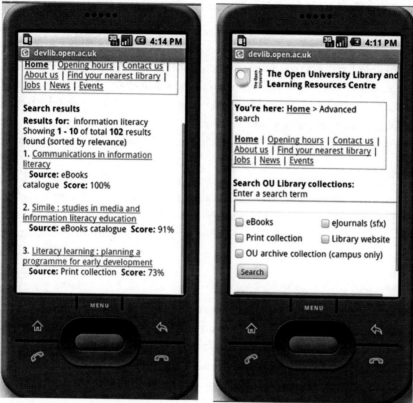

Figure 9.3 The front end of the OU Library consolidated search

As part of our mobile services development project, we have been actively working in the area of information literacy to deliver learning activities to smaller screens. Our e-learning specialist, Anne Hewling, has designed a content model for instructional designers to author mini learning objects that can have a maximum of five pages with no more than 90 words per page. These mini learning objects render well on smaller screens, especially on lower-end mobile phones. This content model has been delivered through our in-house developed Learning Object Generator (LOG) system. With the LOG system, authors can either preview the authored learning objects as web pages or package the contents of those learning objects according to the IMS content packaging standard for deployment directly into Moodle.

Conclusion and further development plans

The landscape for the use of mobile phones is changing rapidly; the sharp increase in the use of Wi-Fi enabled phones and PDAs is making it cheaper for users to stay online; and mobile web browsing is becoming increasingly popular. With the increasing number of mobile users going online, the demand for m-library services is growing as well, which puts pressure on universities and libraries to improve their internet-based systems and develop services which are fit-to-purpose for the smaller screens.

The development teams at the OU and AU libraries have been continuing to collaborate on the development of m-libraries and to deliver content to students via mobile devices. This partnership has played a significant part in helping the OU Library to plan the development of several new mobile library services in the coming months. These plans include:

- Development of a dedicated version of the OU Library website for iPhones and Androids, to take advantage of the advanced features of these phones such as bigger screen sizes, intuitive user interface and good multimedia quality. The work will also include redeveloping the page templates and enhancing the existing style-sheets to work on iPhones.
- Implementation of a web service (in collaboration with AU Library) which will return the profiles (containing details about screen size, make and model, platform and the features supported) of mobile phones that users are connecting from.
- Development of SMS (Short Messaging Service) alerts to notify library users about their reserved items and overdue loans and to offer renewals through their mobile phones.
- Investigation of implementation of SMS reference services as an additional communication channel for library users needing help. A number of academic libraries in the USA are already offering SMS reference services and finding that they have good take-up amongst their users.
- Continuation of work on refining the library search and improvement of our consolidated search prototype.
- Exploration of the possibility of developing a mobile self-assessment tool for the iKnow materials.

- Continuation of work with the OU's Learning and Innovation Office to contribute towards the development of the mobile VLE system. The primary aim of the mobile VLE project is to increase the access to the OU VLE so as to better support students who are already trying to access OU content and services via mobile devices. A longer-term aim of the project is to work with the course teams to investigate and identify parts of their activities and resources that are most appropriate for mobile learning.
- Continued collaboration on mobile library services development with AU and building new collaborations with other interested universities such as Cambridge, UK, Ryerson University, Canada and University of Catalonia, Spain.

Notes

1   Google Analytics,
    www.google.com/analytics/
2   iKnow,
    http://iknow.open.ac.uk
3   IMS Content Packaging Specification,
    www.imsglobal.org/content/packaging/
4   Skweezer,
    www.skweezer.net

References

Hadar, A. (2008) BMO: Apple will sell 26 million iphones in 2010, '09 will be tough, MacBlogz,
    www.macblogz.com/2008/11/06/bmo-apple-will-sell-26-million-iphones-in-2010-09-will-be-tough/

Mills, K. (2009) *M-Libraries: information on the move.* A report produced as part of Arcadia Programme based at Cambridge University Library.

Pettit, J. and Kukulska-Hulme, A. (2007) Going with the Grain: mobile devices in practice, *Australasian Journal of Educational Technology*, 23 (1), 17–33.

10

Designing a mobile device automatic detector to support mobile library systems

Yang Guangbing, Tony Tin, Colin Elliott, Maureen Hutchison and Rory McGreal

Abstract

Content providers cannot ensure that digital material will be reformatted and accessed by any mobile device correctly, due to mobile device limitations. A *mobile device automatic detector* can provide accurate information about devices accessing mobile library systems. This information can be used to properly render content for the specific device. This chapter proposes an approach to designing a detector that will support mobile library systems so as to render web content dynamically and adaptively. The overall architecture of the detector is also discussed in this chapter and follows from a simple experimental study to evaluate the design proposed here.

Introduction

The rapid adoption of mobile devices with internet capabilities has allowed users to work or study at any time, in any place (Motiwalla, 2007). There have also been descriptions of implementations of library access via mobile devices (Needham and Ally, 2008; Yang et al., 2006). However, there are many limitations to mobile devices, which greatly restrict the relevant applications of mobile technology (Ally et al., 2006; Kojiri et al., 2007). Although some content providers have designed purposely digital material for mobile learning, providers cannot ensure that content is able to be correctly reformatted and accessed by any mobile device. This is due to the diverse characteristics among devices, which are not taken into account by most ubiquitous learning systems (Yang, 2007; Motiwalla,

2007). Some early attempts made effective use of proxy servers (Cheung et al., 2007)

In any case, either with proxies or by other detection methods, it is essential to provide adaptive content based on the characteristics of mobile learners and mobile devices. In this chapter an automatic mobile device detector for mobile library systems is proposed to support content designers. This detector will provide adaptive content based on the capabilities of different mobile devices. This investigation concerns and is limited to the characteristics of the mobile devices. Learners' characteristics may also affect adaptive content.

The aim of this research is to construct an architecture that detects features of mobile devices, create RDFS (Resource Description Framework Schema) formatted mobile device profiles and provide content designers with services that are relevant to the mobile device profile. Constructing the mobile device profile is a great challenge for mobile library systems, the most difficult part being the immediate and accurate collection of mobile device features when content is accessed via the world wide web.

Mobile device issues

In order to understand the requirements necessary to develop the mobile device automatic detector, the following issues must be addressed. First, what mobile device capabilities determine the kind of the content that can be displayed correctly? Second, in what context can material be processed properly? Third, can mobile library systems detect mobile device capabilities before preparing and transporting suitable content to mobile clients?

Correctly displaying content

First, although mobile-device optimized web content can provide mobile learners with educational experiences using the mobile library, the limitations of this approach are significant because of the diverse capabilities of the different mobile devices. These divergent characteristics affect or eventually determine which types of web content can be rendered legibly on these devices. Therefore it is important to take these different device capabilities into account in the mobile library system. If the mobile library system can identify these capabilities, it can adjust the content layout, font size and other attributes of a web page, or even convert the web content to different formats at run time. Detailed discussions of the capabilities of the mobile devices which determine the types of web content will be described in the next section.

Detecting mobile device context

Second, the context-sensitive data that are relevant to mobile devices need to be taken into account during web content delivery. For example, if the mobile library system detects the device it is accessing through a wireless local network using Wi-Fi, it may recognize that it is possible to transport more data to the device because the connection speed is fast, the bandwidth is wide and the connection is free of charge. On the other hand, if the system knows that the device is accessing the services via a cellular network using an expensive Telco carrier data plan, it understands that data-intensive web content (e.g. video) could be very expensive for the user. For this reason, it is important that the system is able to detect the context of the mobile devices in order to provide suitable content from the mobile library.

Detecting device capabilities

Third, currently there are several ways to help mobile information systems detect the capabilities of different devices. Wireless Universal Resource FiLe[1] (WURFL) obtains device information from the User-Agent. All of the information contained in the WURFL is organized based on this User-Agent and is structured by families of devices. But WURFL is one of the idealized representations of 'offline' approaches. The system has to store the device's capabilities first. It has no way of acquiring device information ad hoc. UAProfile (WAG UAProf[2]) is another standard concerned with capturing capability and preference information of wireless devices. This information can usually be used by content providers to produce content in an appropriate format for a specific mobile device. A UAProf describes the capabilities of a mobile device, including vendor, model, screen size, multimedia capabilities, character set support and more. A mobile device sends a header within an HTTP request, containing the URL, to its UAProf.

However, relying solely on UAProf has the following drawbacks:

1 Not all devices have UAProfs.
2 Not all advertised UAProfs are available.
3 Retrieving and parsing UAProfs in real-time is slow and can add substantial overhead to any given web request.
4 The UAProf document itself does not contain the user agents of the devices.
5 There is no industry-wide data quality standard for the data within each field in a UAProf. (see Glover and Davies, 2005)

Our approach to this problem is to combine real-time derived information from the HTTP accept header, component analysis, User-Agent data and UAProfiles to deal with the actual device itself rather than the idealized representation of 'offline' approaches such as UAProf or WURLF.

The remainder of the chapter is structured as follows. A further discussion of the above three issues is provided, to identify what mobile device capabilities should be taken into account when providing content and how mobile library systems recognize these capabilities. Following this, an introduction to the overall system architecture of the mobile device automatic detector encompasses the types of components involved in the mobile device automatic detector, its overall architecture and technologies used in the development of the detector. An experimental study to evaluate the paper's proposed approach is provided, followed by a conclusion and further work for study.

Discussion

A mobile device's capabilities are determined by a set of information representing the features of the device. This information includes mobile browser capabilities and product information: e.g., brand name, model name, manufacturer, operating system, processor type and memory size, screen size and resolution, video and audio features, storage, display, image format support, Flash, PDF, WAP, and such. Different mobile devices will have different capabilities.

The following mobile device capabilities are essential to the operation of mobile library systems.

Browser enabled: Mobile library systems know information about the device browser (e.g. Openwave, Nokia, Opera, Access, Teleca) and features of the device. Knowledge of these features is most important when a user accesses mobile library systems from the internet. For example, the Opera browser in Nokia's Symbian operating system (OS) claims web support is available and can render a web page independently from the device. The device accessing the system may not be known, but if the exact type of mobile browser installed on the device is known, then suitable web content can be prepared.

Product information: This includes the brand name, model, type of OS, wireless device web support feature (which describes if a page was developed for web presentation independently from the device) and pointing method (such as joystick, stylus, touch screen, click wheel, etc.), which are also important for web content rendering.

Mark-up language support: Includes language such as WML, HTML, XHTML, CHTML (Compact HTML, iMode browser), CSS and Ajax.

Display information: Includes the screen width and height resolution, number of columns presented in the mobile device screen, number of lines presented in the screen and maximum image width and height.

Image support information: Includes types of images supported by the mobile device, such as GIF, JPEG, PNG, TIFF, BMP, WBMP, etc.

Sound format support information: Identifies what types of sound formatting the mobile device supports (e.g. most mobile devices support WAV and MP3 sound formats).

Other useful capabilities: Includes audio, video, playback and streaming, and J2ME support information.

The mobile library system can prepare and provide suitable web content to a mobile client when it knows what mobile device is being used. At the same time, the mobile device can easily render content, present web pages with support for multimedia (such as video, audio, or Flash clips), and select an appropriate screen layout. For example, a web page can be laid out based on the exact width and height of the resolution of the screen of the mobile device requesting content. This method avoids unnecessary scroll bars being displayed on the screen, especially the horizontal scroll bar, which can discourage the reader or even render the content unreadable.

Overall system architecture

The approach to the architecture of the mobile device automatic detector is service oriented such that the architecture automatically provides mobile device capability detection services. There are three components to the architecture.

The main component is the *Mobile Device Detection Adaptor*. This adaptor determines what content and context formats a device will support, by using any combination of device profile information, such as the HTTP User-Agent header and HTTP Accept headers and UAProf (Mobile Web Best Practices 1.0[3]). The reason for using the combination of the User-Agent, HTTP Accept headers and UAProf (User Agent Profile) is that user-agent headers do not always uniquely identify the device; some devices do not supply accept headers, some misstate their capabilities and UAProf information may be unavailable or incomplete. Thus, a combination of User-Agent, HTTP Accept headers and UAProf can provide the most accurate information about the device.

Another component of this architecture is the *Mobile Device Profile Services Provider*, its main task being to generate RDFS-formatted device profiles. These profiles can be used by mobile library systems or other applications to consume information about devices accessing the system. The third component is the *Device Profile Data Repository*. The raw data of the mobile devices and formatted RDFS device profiles are stored in the repository. Figure 10.1 illustrates the overall architecture of the mobile device automatic detector.

Experimental study

This experimental study focuses on evaluating the design of the mobile device automatic detector. One of the study's aims is to evaluate the compatibility of the detector by testing many mobile devices, from low-end mobile phones to high-end smart devices. Another purpose of the experiment is to assess the accuracy of detection by observing the differences in the presentation of the page, content and context format, and navigation styles and, in particular, by investigating the difference between rendering the same content in a 'true' browser and in a 'mobile' browser when using the same mobile device that supports the 'true' browser (such as the Nokia N810 PDA).

In this study, several brands of mobile devices were tested. They

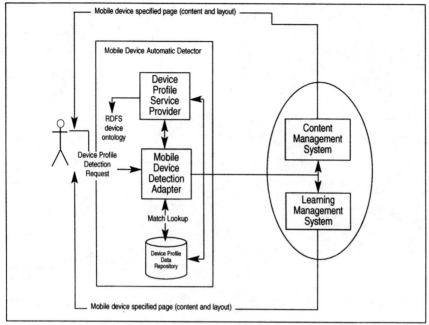

Figure 10.1 Overall architecture of the mobile device automatic detector

included a low-end mobile phone (Samsung SGHD807[4]) which supports WAP, XHTML level 3, XHTMLMP 1.0 mark-up content and images with less than 65,536 colours. Other experiments used an HP iPAQ with Windows Mobile 5 and Microsoft Pocket IE 4, an HTC smart phone P3700 with Windows Mobile 6 and an iPhone with Max OS X. Figure 10.2 illustrates the testing results.

Table 10.1 (on page 105) lists device features detected by the detector and corresponding capability values in the experimental study.

Only some of the device features and capability values are listed in Table 10.1. For more detailed information about the devices (e.g. HTC Touch Diamond), the manufacturer's URL of the UAProfile is provided at the end of this chapter. Since not all mobile devices have the manufacturer UAProf information available, this detector architecture includes a feature in the *Device Profile Service Provider* module to generate the device profile on demand, if no manufacturer-provided or previously generated UAProfile is found in the device profile data repository. The generated sample profile is shown in Figure 10.3. This RDFS document is created based on the standard of CC/PP Processing 1.0 (JSR-000188 CC/PP Processing final release[6]).

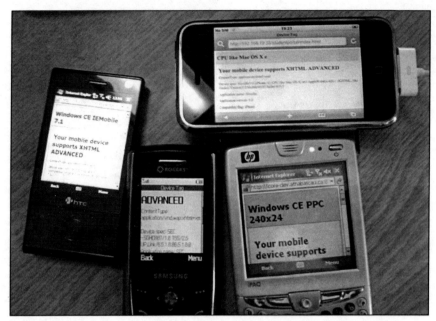

Figure 10.2 Testing results of the mobile device automatic detector using various mobile devices

```
<rdf:RDF>
<rdf:Description rdf:ID='Profile'>
<prf:component>
<rdf:Description rdf:ID='HardwarePlatform'>
<rdf:type rdf:resource='www.openmobilealliance.org/tech/profiles/UAPROF/
    ccppschema-20021212#HardwarePlatform'/>
<prf:Model>iPhone</prf:Model>
<prf:Vendor>Apple Computer Corporation</prf:Vendor>
<prf:BitsPerPixel>16</prf:BitsPerPixel>
<prf:ColorCapable>Yes</prf:ColorCapable>
<prf:ScreenSize>320x480</prf:ScreenSize>
<prf:ImageCapable>Yes</prf:ImageCapable>
<prf:PixelAspectRatio>1x1</prf:PixelAspectRatio>
<prf:ScreenSizeChar>32x48</prf:ScreenSizeChar>
<prf:StandardFontProportional>Yes</prf:StandardFontProportional>
<prf:SoundOutputCapable>Yes</prf:SoundOutputCapable>
<prf:TextInputCapable>Yes</prf:TextInputCapable>
<prf:VoiceInputCapable>Yes</prf:VoiceInputCapable>
<prf:InputCharSet>
<rdf:Bag>
<rdf:li>US-ASCII</rdf:li>
<rdf:li>UTF-8</rdf:li>
<rdf:li>UTF-16</rdf:li>
<rdf:li>ISO-10646-UCS-2</rdf:li>
</rdf:Bag>
</prf:InputCharSet>
<prf:Keyboard>No</prf:Keyboard>
```

Figure 10.3 Sample UAProfile created during the experimental study

Table 10.1 Device features and capabilities tested in the experimental study				
Features	Device model	Samsung SGHD807	HP iPAQ	HTC Touch Diamond P3700	iPhone
	Capability value				
Display	Columns	17	18	16	
	Rows	6	10	36	
	Resolution. width	176	240	520	320
	Resolution. height	220	240	640	480
	Max.image. width	160	220	440	320
	Max.image. height	200	220	480	360
Mark-up support	Html_wi_oma _xhtmlmp 1.0, wml 1.3, xhtml level 3	WML 1.1, XHTML level 1 support	WML 1.1, XHTML, WAP, HTML 4.1	XHTML, accept third party cookie	
Image format	JPG, GIF, BMP, WBMP, PNG, colour:65536	JPG, GIF. Colour:65536	JPG, GIF, BMP, WBMP, PNG.		
J2ME support	MIDP 1.0, 2.0, CLDC1.0, 1.1	MIDP 1.0, 2.0, CLDC1.0, 1.1	MIDP 1.0, 2.0, CLDC1.0, 1.1	No	
Mobile browser	TSS 2.5	MSIE 4.0	Opera 9.5 and Microsoft Pocket IE 4	Safari	
Pointing method				Touch screen	
Manufact-urer provides UAProfile	Yes (Samsung SGHD807 UAProfile)	No	Yes (HTC Touch Diamond P3700 UAProfile[5])	No	

As previously mentioned, the capabilities of various mobile devices are quite different. For example, the number of rows of text that can be displayed using the Samsung SGDH807 screen is 6, but using the HTC device it can be 36. The number of text characters, including punctuation and spaces, that can be shown in one screen using the Samsung mobile device is 102 (6 rows multiplied by 17 columns) whereas for the HTC smart phone this value is 576, five times more than the Samsung phone. If no mobile device detector is implemented in mobile library systems, the system will not know how many characters can be displayed on one page. If more characters are allowed on a page than a mobile device will allow (e.g. the Samsung SGDH807), then the user will have difficulty reading and navigating the content, since the keypad must be used to scroll the page up and down or left and right. If, on the other hand, fewer words are allowed on a page, then the HTC smart phone user may feel uninspired or disengaged, viewing only several words per page. The benefits of implementing a mobile device automatic detector in mobile library systems are therefore clear.

Conclusion and further work

In this chapter, an overall architecture of the mobile device automatic detector has been described and an argument presented on the need for such a detector for mobile library systems, along with a discussion of the features. This research will benefit developers and researchers, since a clear picture has been generated of how great the differences can be among various mobile devices, and how additional value can be provided if systems can detect the myriad features and capabilities of mobile devices before web content is prepared for them. There are some remaining challenges in the development of the detector, however. New technologies are being investigated to develop methods to make the detection processes more accurate and to easily update the device profile data repository. A new approach to speed detection processing is also being explored.

For future work, there is a plan to design and implement web services to facilitate the use of the mobile device automatic detector. In addition, scheduling mechanisms and interface applications to make the operation of the device profile data repository update more easily will be implemented.

Notes

1   Wireless Universal Resource FiLe (WURFL) [accessed 12 October 2009], http://developer.openwave.com/dvl/tools_and_sdk/wurfl_and_wall/

2   WAG UAProf. Wireless Application Protocol, WAP-248-UAPROF-20011020-a
      [accessed 12 October 2009],
      www.openmobilealliance.org/tech/affiliates/wap/wap-248-uaprof-20011020-a.pdf
3   Mobile Web Best Practices 1.0 [accessed 8 September 2009],
      www.w3.org/TR/mobile-bp/
4   Samsung SGHD807 UAProfile [accessed 11 September 2009],
      http://wap.samsungmobile.com/uaprof/d807_10.xml
5   HTC Touch Diamond P3700 UAProfile [accessed 11 September 2009],
      www.htcmms.com.tw/gen/diamond-1.0.xml
6   JSR-000188 CC/PP Processing final release [accessed 11 September 2009],
      http://jcp.org/aboutJava/communityprocess/final/jsr188/index.html

References

Ally, M., Koole, M. and McGreal, R. (2006) Usability of Mobile Devices and
    Designing for Mobile Learning. Paper presented at the International
    Conference on Mobile Communications and Learning, Mauritius.

Cheung, B., McGreal, R. and Tin, T. (2007) *Implementation of Mobile Learning Using
    Smart Phones at an Open University: from stylesheets to proxies.* Paper presented at
    the IADIS Mlearn 2007 Conference, Lisbon, Portugal.

Glover, T. and Davies, J. (2005) Integrating Device Independence and User Profiles
    on the Web, *BT Technology Journal*, 239–48.

Kojiri, T., Tanaka, Y. and Watanade, T. (2007) Device-independent Learning
    Contents Management in Ubiquitous Learning Environment. In: Bastiaens, T.
    and Carliner, S. (eds), *Proceedings of World Conference on E-Learning in Corporate,
    Government, Healthcare, and Higher Education 2007*, Chesapeake, VA: AACE,
    991–6.

Motiwalla, L. F. (2007) Mobile Learning: a framework and evaluation, *Computers and
    Education*, 49 (3), 581–96.

Needham, G. and Ally, M. (eds) (2008) *M-Libraries: libraries on the move to provide
    virtual access*, Facet Publishing.

Yang, M. (2007) An Adaptive Framework for Aggregating Mobile Learning
    Materials, *The Seventh IEEE International Conference on Advanced Learning
    Technologies (ICALT 2007)*, 180–2.

Yang, C., Tin, T., McGreal, R., Ally, M. and Coffey, S. (2006) The Athabasca
    University Mobile Library Project: increasing the boundaries of anytime and
    anywhere learning for students, *Proceedings of the IWCMC 2006*: 1289–94.

11

The Athabasca University Library Digital Reading Room: an iPhone prototype implementation

Rory McGreal, Hongxing Geng, Tony Tin and Darren James Harkness

Introduction

The iPhone, with its ability to support various text and multimedia formats, provides a unique opportunity for libraries to open up access to their digital collections. At Athabasca University (AU) researchers have initiated a process for the implementation of the AU Digital Reading Room (DRR). The DRR[1] was one of Canada's first digital libraries to open up access to library materials on mobile devices.

The iPhone presently represents the state of the art in mobile computing. With its touch screen, novelty of design, broadband access, visual display and multimedia capabilities, it offers enhanced possibilities for facilitating learning. As a first step in the deployment of any iPhone app, even for testing purposes, AU joined in the iPhone Developer Program.[2]

The DRR is an online course reserve repository that provides service both to AU students (providing accessibility) and to the university and its various centres (providing protection). It comprises many digital reading files filled with course readings and other supplementary course-related content. These are faculty-chosen learning resources, housed in a repository in various formats, including learning objects, e-books, e-journals, audio and video clips, websites and book chapters. The available resources have been organized by course and by lesson for the convenience of students and they provide learners with easy access to course materials via PCs and mobile devices. The DRR currently supports 251 online courses, with links to more than 24,000 online resources.

Literature review

Waycott and Kukulska-Hulme (2003) focused exclusively on students' experiences with reading course materials and taking notes using mobile devices. They found that students were able to make notes only with great difficulty. However, this early investigation was conducted using a relatively affordable first-generation mobile device with limited capabilities. According to Clyde (2004), the challenge was 'to identify the forms of education and training for which Mlearning is particularly appropriate, the potential students who most need it and the best strategies for delivering mobile education' (46). Lippincott (2008), Cheeseman and Jackson (2009) and Ally et al. (2008; 2009) have all focused on delivering m-library services to the next generation of students, recognizing the social changes being wrought by the ubiquity of mobile devices. Researchers have already begun investigations into the use of iPods (Coombs, 2009), and now of iPhones, in providing library services (Sierra and Wust, 2009).

Project implementation

The DRR was originally designed primarily for viewing text on a standard PC and CRT monitor using HTML tables. Content viewing on mobile devices was enabled by removing the width-specific code in order to render the layout 'fluid'. In this way, the content can be displayed on a very small screen without the need for horizontal scrolling. Block elements flow to the next line if the horizontal space is lacking.

There are several requirements to create an iPhone application. Currently, the only supported platform for iPhone development is Mac OS X 10.5.8 or above. Likewise, the only language supported for iPhone development is Objective C 2.0 with Apple's iPhone SDK 3.0 or above. iPhone development also imposes restrictions on supported video and audio formats; h264, Quicktime, MP3 and AAC are supported on the iPhone platform, for example, while DivX, Windows Media and Ogg Vorbis are not.

An initial challenge was content delivery. The DRR contains a large amount of content, including audio and video files. Bundling this content into the iPhone application was not an option, given the storage requirements and need to update. As a result, we needed to use a model that allowed us to access data directly from the DRR over a wireless connection.

To meet this challenge, we implemented a client–server architecture. A proof-of-concept client application was created that runs on an iPhone or iPod Touch and accesses the DRR over the internet, making use of XML files stored on the DRR server. For the proof-of-concept client, we created XML files manually on the DRR server. In our next phase of development, these XML files will be created dynamically on the DRR server, based on requests from the DRR iPhone application (other mobile applications making use of the XML API).

The client acts in a similar fashion to a browser, making requests to the server and retrieving XML data in response. This data is then rendered natively in the application. The XML files on the DRR server contain several different content types, which the DRR iPhone application parses and renders appropriately. The following is a short example of a DRR XML file:

```
<?xml version='1.0' encoding='UTF-8'?>
<items>
<item name='example' type='parent' href='example.xml'/>
<item name='type example' type='type' href='text.xml'/>
<itemname='audio example' type='audio'
  href='www.example.com/audio.xml'/>
<item name='video example' type='video'
  href='www.example.com/video.xml'/>
<item name='link example' type='link' author='Example'
  href='www.example.com'/>
</items>
```

The DRR XML files make use of a custom XML schema, with a root element of <items>. Each item is marked with the <item> element and contains attributes to pass along the item's name, type, location in the DRR and author, if applicable. The name element is displayed in the iPhone DRR application as a row in a table, along with the author attribute if present (see Figure 11.1). The type attribute can be set to the following content types:

- *parent* contains other items as its content, which is contained in an XML file
- *text* is plain text content, which is contained in an XML file

- ■ *audio* is an audio file
- ■ *video* is a video file
- ■ *link* provides a link to an external http resource (e.g. website or a PDF file).

Figure 11.1 iPhone Digital Reading Room example

The type attribute works in concert with the href attribute to direct the iPhone DRR application. The following are examples of how the iPhone DRR application would handle certain content types:

- ■ *type*='*parent*', *href* points to an XML file containing more items, which are shown in a table
- ■ *type*='*text*', *href* points to an XML file, which contains an item holding text in name attribute
- ■ *type*='*audio*', *href* will be a link which points to an audio file
- ■ *type*='*video*', *href* will be a link which points to a video file
- ■ *type*='*link*', *href* will be an external link.

Through configuration like *type*='*parent*', contents are organized hierarchically.

Each item in an XML file is rendered as a row in a table, using the iPhone SDK's standard table view (UITableView). Items with a type of 'parent' link to additional XML files on the server and render as a new table. Text content is displayed in a standard text area (UITextView). Audio and video content is played natively in the iPhone OS's movie player (an instance of MPMoviePlayerController). An integrated instance of Mobile Safari (UIWebView) is used to display externally linked web content to users.

Conclusion

This project was designed to test the boundaries of the educational use of the latest iPhone technology, specifically by enabling and increasing accessibility to the m-library. Building on the research conducted in Canada and elsewhere, the researchers are investigating other mobile devices and the use of international standards in order to maximize the potential of m-libraries.

From a developer's perspective, the research has provided us with experience in determining modalities for such an implementation. It also provides guidance for other researchers as they investigate the shifting terrain of mobile learning. To summarize, a mobile delivery application has been set up in the DRR enabling access by iPhones. This device has been used to access the varied resources in the mobile version of the DRR. The researchers have compared the PDAs and identified some of the critical features of each operating system and pocket browser.

In order to view a prototype of the DRRIPHone implementation, please go to http://library.athabascau.ca/drr/iphone/drr.zip.

Notes

1   http://library.athabascau.ca/drr
2   Apple.com iPhone University Developer Program,
      http://developer.apple.com/iphone/program/university.html

References

Ally, M., Elliott, C., Schafer, S. and Tin, T. (2009) Mobile Library: connecting new generations of learners to the library in the mobile age. Paper presented at the Second Annual M-Libraries Conference, Vancouver.

Ally, M., McGreal, R., Schafer, S., Tin, T. and Cheung, B. (2008) Use of a Mobile Digital Library for Mobile Learning. In: Needham, G. and Ally, M. (eds), *M-libraries: libraries on the move to provide virtual access*, Facet Publishing, 217–27.

Cheeseman, P. and Jackson, F. (2009) Bridging the Mobile Divide: using mobile devices to engage X and Y generations. Paper presented at the Second Annual M-Libraries Conference, Vancouver.

Clyde, L. A. (2004) M-Learning, *Teacher Librarian*, www.teacherlibrarian.com/tltoolkit/info_tech/info_tech_32_1.html

Coombs, K. (2009) Piloting Mobile Services. Paper presented at the Second Annual M-Libraries Conference, Vancouver.

Lippincott, J. (2008) Libraries and Net Gen Learners: current and future challenges in the mobile society. In: Needham, G. and Ally, M. (eds), *M-Libraries: libraries on the move to provide virtual access*, Facet Publishing, 17–23.

Sierra, T. and Wust, M. (2009) Enabling Discovery of Digital Collections on Mobile Devices. Paper presented at the Second Annual M-Libraries Conference, Vancouver.

Waycott, J. and Kukulska-Hulme, A. (2003) Students' Experiences with PDAs for Reading Course Materials, *Personal and Ubiquitous Computing*, 7 (1), 30–43.

Part 3
Application of m-libraries

12

Mobile access for workplace and language training

Mohamed Ally, Tracey Woodburn, Tony Tin and Colin Elliott

Introduction

The 'Workplace English' site was created for English language learners to study grammar and vocabulary that can be used every day at work. Vocabulary and workplace-specific situations are used to teach useful, everyday English. The site was designed to be accessed by mobile devices, allowing learners to choose their own classroom and their own study schedule, e.g. on the bus or in their lunch break at work. The site created in this project can be duplicated for other library applications that will allow learners to access information and course materials at any time and from anywhere. The information and course materials are stored in the library repository.

Athabasca University (AU) created the Mobile ESL (English as a second language) site in 2007 (www.eslau.ca). It was pilot tested with several ESL groups, with successful results. The students found the site to be useful, but wanted more listening practice, and situations that they could watch and use as models for functioning in an English-speaking world. Students also asked to see vocabulary that they might face in workplace situations. This resulted in the creation of the Athabasca University Workplace English site, www.wpeau.ca (see Figure 12.1).

The course content consists of over 80 lessons and related exercises teaching the basics of the English language, ranging from the difference between 'is' and 'are' to verb tenses, countable nouns and other aspects of basic grammar. These digital lessons have been adapted into reusable multimedia learning objects which are accessible to anyone on the internet

Figure 12.1 Athabasca University Workplace English site

either as stand-alone lessons, as groups of lessons in units or as full course modules. The content materials are in digitized form, with interactive elements added to enhance flow and motivation.

Specifically, the content has been rendered interactive using a variety of multiple-choice, short-answer, jumbled-sentence, matching/ ordering or filling-in-the-gap exercises on the world wide web and it has been specifically formatted for output using small mobile devices.

Explanation of content

Before creating some of the content for this site AU spoke with ESL students living and working in Northern Alberta. They asked to see content that reflected their life situations more clearly and to learn vocabulary they were hearing everyday in the workplace. Specialized vocabulary from eight different career fields was added to typical grammar lessons. Instead of using the present perfect formation with 'I have been learning English for four years', worksite-specific vocabulary was added to form sentences such as 'I have been welding on the pipeline for two months.' The eight themed fields are:

1 Health and Wellness
2 Utilities-Oil/Gas/Mining
3 Retail and Sales
4 Hospitality and Tourism
5 Food Services and Restaurant
6 Trades and Labour
7 Administration and Clerical
8 Banking and Personal Finance

Vocabulary from these areas was chosen after researching the job openings in Northern Alberta at the time that the site was created (2008). These fields had multiple openings and seemed broad enough areas to employ a large number of people, while at the same time having specific vocabularies that someone learning English might not have encountered in a typical grammar textbook.

At the beginning of each themed section vocabulary lists were added containing words that would be encountered in the upcoming lessons and exercises. The 'Health and Wellness' list was developed as a listening exercise as well, with each word from the list being 'clickable'. By clicking on a vocabulary word in the list a learner can listen to it as a listening and pronunciation exercise. Interactive audio and video lessons on the site provide students with a chance to see and hear situations in the workplace. For example, a dialogue between a waiter and a manager contains a realistic conversation that might happen on an everyday basis in a restaurant. Or a video scenario takes place in a clinic after an on-the-job injury. The vocabulary includes words and phrases such as: 'Do you have any allergies?' and 'incident report', words and questions that it is necessary to know in this kind of situation. After the Workplace English site content was created, the next step was to create the 'mobile' part of the project.

Technical needs

The iPhone represents a large change in the capabilities of mobile devices. Despite the great improvement, technical difficulties remain. The first challenge is formatting the content for mobile devices. An iPhone can be used to view complete non-mobile web pages, but it is preferable to make navigation simple and to keep the content for each page short. Small, easily navigable pages are ideal. For media used to enhance the lessons, there are specific considerations for both format and formatting. Images should be small and fit on the screen. Large images should be avoided when possible because they force users to scroll unnecessarily. For video, the MP4 file format should be used. Other common formats such as .AVI or .WMV cannot be played on the current iPhone model. A balance of compression and quality needs to be maintained so as to ensure that the video is not too large to be loaded quickly onto the device and that the quality is still good enough for the user. Audio requirements for the iPhone are simpler, as the common MP3 file format works easily. With these challenges and guidelines in mind, it is much easier to ensure that users receive the desired content and have a positive and full experience.

Pilot testing results

Adult learners assisted AU with pilot testing for this project. The student volunteers were taking courses in English as part of their course load, or had taken English courses recently. All student feedback reported here is taken from the project report, *WPEAU Mobile Learning Data* (2009). Seventy-nine percent of the students were between the ages of 18 and 25 years old. Fifty-seven percent of the students had never used a mobile device before, while 36% used mobile devices (Blackberry, IG) on a daily basis.

Participants were trained to use the iPhone to access the Workplace English materials. The training session consisted of slides demonstrating the different question types and how to use the functions needed during the testing (bookmarks, back button, submit, etc.) Also included was a live demonstration of the zoom function on the iPhone. The students used a first-generation iPhone to complete the workplace learning lessons. Following the lessons, they completed an extensive questionnaire which included agreeing (or disagreeing) with the statements that follow.

The technology provides flexibility for me to learn anywhere and at any time

Eighty-six percent either agreed or strongly agreed that this technology would provide them with the flexibility to learn in their chosen location at their leisure (Table 12.1).

Table 12.1 The technology provides flexibility for me to learn anywhere and at any time

Level of agreement	Percentage
Strongly agree	43
Agree	43
Neutral	14
Disagree	0
Strongly disagree	0

One student wrote, 'The way the technology is set up, I can learn anywhere, anytime. This seems very realistic, and the future of learning' (Student L.).

Of the students who did not agree (or were neutral) they stated that they would rather 'learn orally' as the [mobile] lessons took 'valuable time to download' (Student H.). As mentioned above, this pilot test used first-generation iPhones (the newest third-generation iPhone uses a 3G network which is 2.4x faster than the first-generation EDGE network). As the students were trying to access audio and video lessons at the same time, the speed of the network, the number of ports available on the server could have been a factor in slowing access to the multimedia links.

Learning with the mobile technology increases my enjoyment of learning

Seventy-nine percent of the students enjoyed the idea of using a mobile device to study (either 'agreed' or 'strongly agreed' with the statement). One student thought that 'New technology always makes people interested. I love the idea of learning online and on mobile' (Student L., quoted in *WPEAU Mobile Learning Data*, 2009).

One student who didn't agree felt that, 'Its [*sic*] nice but I would rather have a computer for learning, the phone would mainly be used for recreational use' (Student I.). This student went on to comment that 'I feel that its [*sic*] just not really as useful as a computer is'. Again, the 'dissenting' comments came combined with the wait times of three minutes or more to access the multimedia lessons.

The use of this type of technology could make learning materials more easily available

One student agreed with this statement, but was worried that the technology might be too expensive (Student I., quoted in *WPEAU Mobile Learning Data*, 2009). The cost of an iPhone, coupled with the service contract, made some of the students feel that this sort of technology was out of their reach. It was explained that the grammar and audio lessons could be accessed on a regular 'cheap' mobile phone, but some still worried about the cost of connecting to the internet with a mobile device in Canada. But overall, 79% of the students agreed with student L., who stated that this technology 'would be useful if you are [*sic*] away from a computer' (Student L., quoted in *WPEAU Mobile Learning Data*, 2009).

The audio helped me learn from the lessons

The students accessed five audio lessons and exercises. Eighty-six percent agreed that the audio helped them learn from the lessons. A comment from one of the 7% who strongly disagreed with this statement was, 'I like "old fashion" learning, by a teacher in a classroom sometimes technology can fail and interrupt [*sic*] learning when we depend on education' (Student H., quoted in *WPEAU Mobile Learning Data*, 2009). A student who did agree that the audio was helpful felt that the audio was 'too fast' (Student K., quoted in *WPEAU Mobile Learning Data*, 2009). The audio lessons, exercises and dialogues were read at a normal speed. In the appendices, transcripts of the audio sections were provided to give students an opportunity to review and clarify the listening portions if necessary.

I would like to take other lessons using mobile technology

Seventy-one percent would like to take other lessons using mobile phones, making comments such as 'I could definitely learn on a phone. I was almost MORE interested using a phone' (Student L., quoted in *WPEAU Mobile Learning Data*, 2009) and 'I really enjoy the new mobile technology iphones. I think they would be really helpful with all English classes. Thanks for all the information it was really helpful' (Student N., quoted in *WPEAU Mobile Learning Data*, 2009). Some of the 29% who were either neutral, disagreed or strongly disagreed with the above statement commented on the speed of the connection: 'My overall time was spent waiting for something to load' (Student K., quoted in *WPEAU Mobile Learning Data*, 2009). Even students who agreed that they would like to take other lessons using mobile technology felt that they 'would recommend this technology, but it is a little slow' (Student E., quoted in *WPEAU Mobile Learning Data*, 2009). The speed of the connection was an unexpected factor in the pilot test. Some of the students were unfamiliar with the device and thought the phones were not working when the video or audio lessons did not come up immediately. Wait times of under one minute seemed to be accepted, but times over that would have the students asking for help or moving on to a different, non-multimedia style of exercise.

I would recommend that other students complete their courses using mobile technology

Overall, 72% would recommend taking a course with this technology to their friends (Table 12.2).

Table 12.2 I recommend other students complete their courses using mobile technology	
Level of agreement	Percentage
Strongly agree	50
Agree	22
Neutral	7
Disagree	7
Strongly disagree	14

One student commented, 'I like the program on the Iphone. It was helpful to me, because I have a problem with spelling and read. I want one of this Iphone' (Student J., quoted in *WPEAU Mobile Learning Data*, 2009). The students who wouldn't recommend this to their friends commented solely on the slow connection speeds. Other positive comments were, 'I really enjoy the new mobile technology iphones. I think they would be really helpful with all English classes' (Student N., quoted in *WPEAU Mobile Learning Data*, 2009).

Summary

The students seemed very pleased with the idea of the technology, but some were disappointed with the delivery. With the use of next-generation phones and a faster connection speed, the pilot testing could have seen more positive comments, as the one consistent negative thread running through the pilot test was the 'wait time' for the audio and video sections. Though many of the students were patient with the loading times, some had the impression that the phone wasn't capable of delivering the lessons to them in a fast and accurate manner. The wait time was observed at less than one minute in some cases, but this was considered to be too long by some of the participating students.

Overall, the students who could access all of the features liked the interactive activities and the quality of the exercises. They were excited by the technology and wanted to see this style of learning integrated into their course of study. Some of the students mentioned that they travelled frequently with one of the college's sports teams and they thought learning with the iPhone would be very suitable for their long periods on the bus travelling to and from games. Students thought that this technology would 'replace teachers', and that this was an agreeable idea for two of the students. They found the phones easy to work with and thought that the 'information was pretty good.'

The teachers involved in the study (as student supervisors) were very interested in the technology and had many ideas about how to apply it to their courses, including sending the iPhones with students during the on-site portions of their training. This would enable the students to access course materials while 'on the job site.'

Positives	*Negatives*
New, exciting technology	Cost of the hardware (phones)
Very flexible for learning at one's own pace	Cost of the internet connection
	Loading time
Could replace a teacher	

Future work

The Workplace English site focused on building grammar skills, teaching vocabulary for the workplace and listening practice. The next step is to build a site where students can practise their spoken English anytime, anywhere, with a mobile device to guide and assist with pronunciation. A pronunciation practice section with an animated phonetic alphabet and interactive exercises created to practise spoken English is the next step for the Workplace English team.

Reference

WPEAU Mobile Learning Data (2009) Unpublished project report.

13

Service models for information therapy services delivered to mobiles

Vahideh Z. Gavgani

Introduction

Mobile computer technology has increasingly penetrated the healthcare industry. The increasing capabilities of mobile phones, along with the powerful wireless network, have given a new dimension to earlier versions of telemedicine, which may be called mobile-health/m-health. Healthcare providers and physicians use mobile phones and the Short Message System (SMS) to exchange healthcare information with their colleagues and clients. Medical book publishers are changing the format of the medical reference books from electronic/online versions into mobile phone-compatible versions.

A recent approach in healthcare and patient safety that has gained significant attention from healthcare policy makers, health libraries and providers of consumer health information is information therapy (Ix), 'the prescription of evidence-based medical information to a specific patient, caregiver, or consumer at just the right time to help the person make a specific health decision or behavior change' (Kemper and Mettler, 2002, 3). Information therapy relies on ubiquitous, reliable and real-time evidence-based information at the point of care. 'Information Therapy can range from an e-mail sent to every plan member with borderline blood pressure to recommending a website to a physician providing a diabetic patient with a copy of an article on the link between diabetic and heart disease' (*Consumer Driven Healthcare*, 2004).

Both physicians and patients/caregivers need to receive the right information at the point of care to be able to make the right shared decision about the health problem. In addition, effective management of chronic diseases, common diseases and the majority of incurable diseases depends on behaviour change on the part of patients. Furthermore, most diseases are preventable and in such cases information is the main remedy. Therefore, correct information at the right time, for the right person, through the right media, that is available at any time and anywhere, seems to be the best solution for a successful information therapy service as well as for patient empowerment.

Although information is freely and widely available through the internet, in developing countries like India and Iran there are geographically isolated places lacking both internet connection and an appropriate health library (Taylor et al., 2001; Golozar, 2007). Therefore readily accessible health information via the internet is not an option for patients and physicians in these isolated areas. This chapter therefore proposes a service model for delivering information therapy service through mobiles, PDA, fax/e-post to improve patient care and ensure health information for all in rural areas. The model is proposed for those areas lacking internet connection; however, it is also applicable for areas that do have internet access.

Background of study

At one time using mobile phones in libraries was controversial and was banned, but now libraries are seriously engaged in expanding their services to mobile phones. During 2000s libraries have undergone a crucial change in techniques and services in order to keep in line with changes in information and communication technology (ICT) and with users' preference for using mobile phones – but without as yet documenting and publishing their services. As a simple example, while most librarians once thought of the mobile phone's ring in the library as 'problem patron behaviour' (LISWIKI, 2009) (according to the library's policy and the patrons' right to a calm and quiet environment for reading and study), at the same time the capabilities of mobile phones were influencing the development of the library's information services.

Since 2000, libraries have been using mobile phones to provide traditional services such as reservations, ready-reference, book renewals and information requests or orders. This author's experience in the Medical School library of Tabriz University of Medical Science and

Education (www.tbzmed.ac.ir) in offering these services via phone or mobile phone could be cited as an example of this.

Furthermore, a review of the relevant literatures indicates that libraries are using mobile phones in their information services in a variety of ways, from informing clients to user education. Osaka Municipal Library uses mobile phone facilities as well as e-mail to inform its users about the availability of their reserved books (Osaka Municipal Library at www.oml.city.osaka.jp). In Finland public libraries use mobile phones for traditional library services such as reservations, loans and renewals via SMS and WAP (Wireless Application Protocol). The library also uses MMS (Multimedia Messaging Service) to direct patrons to the library through the Library Map and instructions (Ulla-Maija, 2006). According to Ulla-Maija, in 2005 the Finland County Administrative Board researched the use of mobile and net services in Finnish libraries on behalf of the Ministry of Education. The research revealed that SMS services were used in 50% of libraries, SMS for reservation purposes in 49% and expiry date reminder and overdue reminders in 14%. It also indicated that only a few libraries offered internet and WAP services via mobile phone.

Athabasca University (AU) Library has taken an active role in advocating mobile learning within the institution and has been developing mobile-friendly resources and services for a diversity of learners since 2004. AU Library has pioneered the development of a mobile digital library, developed and promoted mobile-formatted resources and services for e-learners using wide ranges of hand-held devices, and engaged in mobile learning content development projects, such as mobile ESL lessons for new immigrants. AU Library has provided strong leadership and an effective support system for mobile learning at the university (Ally et al., 2007a). Athabasca University (AU) has also implemented a comprehensive m-library website. The m-library system can auto-detect users' devices and direct them to the appropriately formatted version (mobile or desktop) of website content (Yang et al., 2008). Ally et al. (2007b) in their presentation entitled 'From M-Library to Mobile ESL: Athabasca University as an Advocate for Mobile Learning', highlight two case studies from Athabasca University: (1) M-Library: Mobile Digital Reading Room Initiative and (2) Mobile ESL: Learning English as a Second Language. They state that 90% of users strongly agree or agree that technology provides flexibility to learn anywhere, anytime and that 60% of users strongly agree or agree that they would take other lessons using mobile technology.

Woody Evans (2006), in an article entitled 'NextGen: Phones Are "Everyware"' introduces Cue Semacode, 'a nifty little application that attaches a URL to a sort of 2-D barcode' that is generated on the website and printed out on sticky labels. These freely downloadable barcodes are readable by mobile phone cameras. Once loaded, they take the user to the designated URL. Library users can capture digital images of 2-D barcodes from printed labels stuck to shelving units, magazine racks, special events displays and even books and go directly to rich online content, tailored to the specific needs of a class, club or reading group.

A review of the literature suggests that there are no specific examples in medical libraries of health information services or information therapy services being delivered to mobile phones. However, there are studies referring to the use of PDAs (Personal Digital Assistants), mobile phones and SMS for medical information and client information by non-librarian bodies. Barrett, Strayer and Schubart (2003), in their study, refer to three common ways in which PDAs are used by residents (doctors studying for speciality medicine), including: (1) medical references (e.g. Five-minute Clinical Consult[1]), (2) pharmaceutical information (such as ePocrates[2]) and (3) professional organization (calendar, address book).

Text messaging or SMS is used by many organizations and healthcare centres to inform clients about appointment times, remind patients about taking medicine, dosage and injection on time. 'Sweet Talk' is a novel intervention designed to support diabetes patients between clinic visits, using text messages sent to a mobile phone. Scheduled messages are tailored to patient profiles and diabetes self-management goals, and generic messages include topical 'newsletters' and anonymous tips from other participants. The system also allows patients to submit data and questions to the diabetes care team (Franklin et al., 2008).

Evidence indicates that surgeons use SMS to deliver instructions in surgery to colleagues without experience or facilities. A BBC report about a vascular surgeon working for Médecins Sans Frontières in the Democratic Republic of Congo who received instructions regarding a life-saving operation via the text messaging/mobile phone from a colleague in the UK is only one example (BBC News, 2008).

Learning about Living (LaL) is a two-year project that uses SMS to teach adolescent reproductive health education to young people in Nigeria (Learning about Living, 2009). SEXINFO is another similar service that was developed by Sexuality Information Services, Inc. in partnership with the San Francisco Department of Public Health. SEXINFO is an information and referral service that can be accessed by texting 'SEXINFO' to a five-digit number from any wireless phone. A consortium of community organizations, religious groups and health agencies assisted with identifying culturally appropriate local referral services (Levine et al., 2008).

Telemedica Support programme VIE-DIAB uses mobile phone and SMS for glycaemic control in adolescents with Type 1 diabetes. Rami et al. (2006) stated that 'During the TM phase, the patients send their data (date, time, blood glucose, carbohydrate intake, insulin dosage) via mobile phone, at least daily, to our server and diabetologists who send back their advice via short message service (SMS) once a week.'

However, a review of relevant literature showed that there is no evidence to indicate medical libraries using mobile phones to introduce Information Therapy services. In remote and rural areas, that have no internet connection (non-wired areas), access to reliable health information and evidence-based information is a critical need for both physician and patients, but it is more expensive and sometimes not possible. The following model is therefore suggested for remote/rural and non-wired areas.

Why information therapy services through mobile phones?

It is believed that a ubiquitous health information service through mobile phone can be the best solution for information therapy services in rural, geographically isolated and non-wired areas. 'With improving capabilities and cheaper rates, mobile telephony is a domain that provides a powerful opportunity for innovation, and because it is subject to lower financial and educational barriers, it provides a potentially wider sphere of influence than the internet has to date' (Anta, El-Wahab and Giuffrida, 2009). 'By the end of 2007, 64% of people in developing countries had mobile phones while in 2002 it was only 44%. But at the same time in developing countries a meagre percentage i.e. 13% were using the Internet' (Lahrodi, 2009). In Iran, SMS texting is currently the largest independent network for exchanging information (TGI, 2009). According to the Cellular Operators Association of India (COAI), the total number of GSM users on 31 March 2009 was 192.35 million, an increase of 4% from February 2009, when the baseline stood at 185 million. The growth in March was 'the highest ever subscriber additions since the inception of GSM service in India' (ThaIndian Portal, 2008).

Furthermore, fax and postal networks with reliable infrastructure are available in most of the rural areas of Iran and India. The Indian postal network has developed its service in hilly, desert and inaccessible areas (India, 2006, 180). 'In addition there are two internet based value added services namely e-post and e-bill post. Through e-post, electronic messages are booked at any post office in India and are downloaded at an identified post office and delivered to the recipient(s) as hard copies thereby connecting individuals who do not have access to personal computer/internet and thus reduce the digital divide. E-Post also offers customers the opportunity to send messages to multiple destinations from a single source.' (India, 2006, 182). This facility raises the possibility of information therapy (Ix) services to inaccessible, rural areas in India.

The model

The suggested model (Figure 13.1) illustrates an information therapy service through mobile phone/PDA and fax/e-post for rural/geographically isolated areas, in India in particular, and in Iran or any other countries with similar facilities and problems in general.

The process of information therapy service delivery through tele-communication technology (mobile phone, fax, PDA, and ePost) include the following steps and procedures:

1 The physician sends the health information request to a librarian through SMS.
2 The librarian receives the request and proceeds to search, retrieve, assess, personalize and finally send the information to the following points:
  a physician's mobile or PDA (in PDF, Symbian or any compatible format where both parties have the devices)
  b post officer through fax/e-post and
  c patient's mobile phone; according to the information prescribed (IP) by patient's doctor.
3 Feedback from the receivers and request for additional information will link the procedures through SMS.

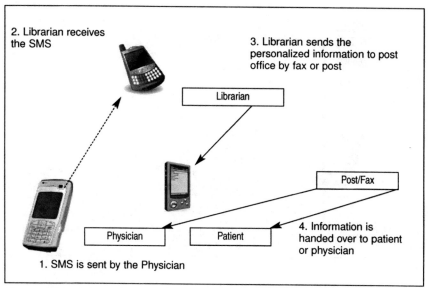

Figure 13.1 Information therapy service using mobile phone/PDA and fax/ e-post for rural/geographically isolated areas

The process and components of the Information Therapy service are presented in Figure 13.2. In this model a physician sends a message (SMS) to the librarian at a health information centre (which may be a national, regional or local health information system or a nearby clinical/medical library). There can be an agreement in the national health information system between the library and clinics/hospitals/doctors in remote and rural areas for provision of evidence-based information (EBI), patient education guidelines, personalized health information/information prescriptions. The librarian receives the message and personalizes the information and immediately sends it by e-post or fax to the post office in the targeted place. The postal officer receives the information, produces a hard copy and hands it over to the physician or patient. Receipt of the information, the need for another piece of information or feedback can also be communicated by SMS. Or the library can send the information directly to physicians/patients via smart phone or PDA if the facility is available.

Some rural clinics are affiliated to medical schools and universities with medical libraries. They can request/receive the information from the same libraries or through them, because most academic libraries have

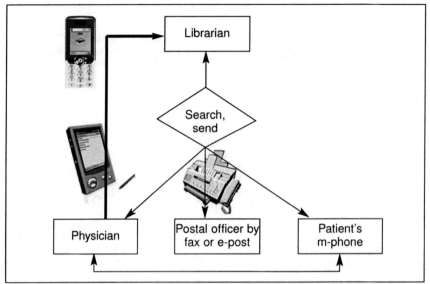

Figure 13.2 Components and process of information delivery using mobile phone, PDA, fax

Note: ——————▶ = need ———▶ = information

agreements with other libraries and information centres, and they usually subscribe to e-journals, consortia and other health information centres. But others are private or affiliated to smaller colleges that may not have access to evidence-based information for financial or other reasons. They can make arrangements with the National Medical Library (in the case of India) or nearest medical library in the province or state (in the case of Iran).

Conclusion

The aim of this model is to identify the overall flow of information and how mobile phones and PDAs might improve the information flow in clinical settings, especially in remote areas where there are no internet access and medical libraries. The transparent, rapid access and ubiquitous nature of mobile technology has made it a desirable means for delivering, transferring and exchanging information.

Librarians have started using SMS for traditional library services, but sending Evidence-based information (EBI) for patients and physicians in the context of information therapy is a new service that can be delivered to distant clients (physicians, patients, caregivers), especially in rural areas, utilizing the ubiquitous, easy-to-use and easy-to-access, comparatively cheap service of mobile phone technology. It is believed that a ubiquitous health information service via mobile phones can be the best solution for information therapy services in rural, geographically isolated and non-wired areas. The socio-cultural, literacy, lifestyle and psychological aspects of the people and community are also important factors that should be considered in delivering such services.

Notes

1   Five-minute Clinical Consult,
     www.studentdoc.com/five-minute-clinical-consult-pda.html.
2   ePocrates, www.epocrates.com/.

References

Ally, M., Schafer, S., Tin, T., Elliott, C. and Lee, S. (2007a) Library as an Advocate of Mobile Learning: the Athabasca University experience [accessed December 2008],
      http://library.open.ac.uk/mLibraries/2007/abstract/lib_adv_moblearn.pdf

Ally, M., Tin, T. and Elliott C. (2007b) From M-Library to Mobile ESL: Athabasca University as an advocate for mobile learning [ppt], [accessed February 2009],
      http://library.open.ac.uk/mLibraries/2007/presentations/Day2_11.40_Tin.ppt

Anta, R., El-Wahab, S. and Giuffrida, A. (2009) *Mobile Health: the potential of mobile technology to bring healthcare to the majority*, IDB Inter-American Development Bank, Innovation Note [accessed March 2009],
      http://idbdocs.iadb.org/wsdocs/getdocument.aspx?docnum=1728061

Barrett, J.R., Strayer, S.M. and Schubart, J.R. (2003) Information Needs of Residents during Inpatient and Outpatient Rotations: identifying effective personal digital assistant applications, AMIA Annual Symposium Proceedings Archive, 784, www.ncbi.nlm.nih.gov/pmc/articlesPMC1480061/.

BBC News (2008) Surgeon Saves Boy's Life by Text, (3 December),
      http://news.bbc.co.uk/1/hi/uk/7761994.stm

Consumer Driven Healthcare (2004) 'Prescription Strength Information: information therapy' fuels wise consumer healthcare choices, *Consumer Driven Healthcare*, 3 (11), 1.

Evans, W. (2006) NextGen: phones are 'everyware', *Library Journal*, (15 July) [accessed March 2009], www.libraryjournal.com/article/CA6349047.html.

Franklin V. L., Greene A., Waller, A., Greene, S. A. and Pagliari, C. (2008) Patients' Engagement with 'Sweet Talk' – A Text Messaging Support System for Young People with Diabetes, *Medical Internet Research*, 10 (2), (Apr–Jun), e20.

Golozar, A. (2007) Rural Doctors, Evidence-based Medicine: the dilemma of best research evidence, *Young Voices in Research for Health*, www.globalforumhealth.org/filesupld/Young%20Voices/07/art/YoungVoices07_Golozar_RuralDoctorsEvidenceBasedMedicine.pdf; www.pubmedcentral.nih.gov/articlerender.fcgi?tool=pubmed&pubmedid=18653444.

India (2006) *India 2006: a reference annual*, compiled and edited by the Research, Reference and Training Division, Publication Division, Ministry of Information and Broadcasting, Government of India. New Delhi.

Kemper, D. W. and Mettler, M. (2002) *Information Therapy: prescribed information as a reimbursable medical service*, Boise, Idaho, Healthwise.

Lahrodi, N. (2009) Iran Stands Lower than its Neighbors in ICT indices, *ITIran*, (10 March (=20 Esfand 1387)), [accessed March 2009], www.itiran.com/?type=article&id=10447.

Learning About Living (LAL) (2009) *Newsletter*, 3 (1), January 2009, [accessed March 2009], www.learningaboutliving.org/mmbase/attachments/25597/Newsletter.

Levine D., McCright, J., Dobkin, L., Woodruff, A. J. and Klausner, J. D. (2008) SEXINFO: a sexual health text messaging service for San Francisco youth, *American Journal of Public Health*, March, 98 (3), 393–5 [accessed March 2009], www.ajph.org/cgi/content/abstract/98/3/393, pre-print version available from: www.ajph.org/cgi/reprint/98/3/393.pdf.

LISWIKI (2009) *Problem Patron Behavior* [accessed March 2009], http://liswiki.org/wiki/Problem_patron.

Rami, B., Popow, C., Horn W., Waldhoer, T. and Schober, E. (2006) *Telemedial Support to Improve Glycemic Control in Adolescents with Type 1 diabetes mellitus*, European Journal of Pediatrics, 165 (10), 701–5.

Taylor, J., Wilkinson, D. and Blue, I. (2001) Towards Evidence-Based General Practice in Rural and Remote Australia: an overview of key issues and a model for practice, *Rural and Remote Health 1* [accessed March 2009], http://rrh.deakin.edu.au.

TGI (2009) Mobile Phone Market Booms in Iran. TGI Survey [accessed March 2009],
   www.tgisurveys.com/news/kmrupdate/KMRUpdate19_2.pdf.
ThaIndian Portal (2008) India Surpasses US as Second Largest Wireless Market, (11
   April), www.thaindian.com/newsportal/business/india-surpasses-us-as-second-
   largest-wireless-market-2_10037056.html [accessed March 2009].
Ulla-Maija, M. (2006) Library Services by Mobile Phone and Internet. NAPLE –
   Conference of the National Authorities on Public Libraries in Europe, (19
   October) [ppt],
   www.naple.info/helsinki/ulla_maija_maunu.pdf.
Yang, C., Tony, T., Rory, M., Ally, M. and Sherry, C. (2008) The Athabasca
   University Mobile Library Project: increasing the boundaries of anytime and
   anywhere learning for students, International Conference on Communications
   and Mobile Computing, *Proceedings of the 2006 International Conference on Wireless
   Communications and Mobile Computing*, Vancouver, British Columbia, Canada
   [accessed February 2009],
   http://hdl.handle.net/2149 /1762.

14

Health literacy and healthy action in the connected age

Paul Nelson and Bob Gann

Introduction

Information technology innovation is opening a wealth of opportunity for educators and health information providers to improve learning, leading to personal and social good. In the case of education, that might be success in course completion and preparedness for professional training. In the case of health education, that might lead to improved individual health literacy, changed attitudes to health and, ideally, health behaviour change. The opportunity is to do learning better, and not only to impart knowledge but also to crack the nut of supporting change for improved health behaviour. New technologies offer vast opportunities to push people into establishing new, healthy behaviours. They also beg new questions, choices and challenges.

- How can we harness the enormous power of the world of information plenty?
- What structures/platforms will deliver it most effectively and safely?
- How can we minimize cost, while bringing along the professionals (teachers, librarians, clinical professionals)?
- How can we ensure that we do good and not harm?

Opening the gates to health knowledge

The technological march underlines an urgency for reflection among knowledge specialists. Prior to the turn of the century, library services gathered knowledge in a world of knowledge scarcity. The mission was the equitable provision of scarce information, learning and thought; the challenges were that provision was severely limited in time and place. In effect, libraries have been the gatekeepers to knowledge, performing rationing in a world of scarcity. In parallel, doctors, particularly primary care clinicians (in the UK) have had the role of gatekeepers to health care and, to a large degree, of gatekeepers to health knowledge. Doctors have been custodians of scarce medical knowledge, both specific to individuals and to general knowledge about illness. The greatest disadvantage of the era of large book and periodical collections was perhaps the inequity of access to knowledge. Inevitably, the less-privileged in society had the least access. The greatest advantage of the large physical collections was perhaps that there was relatively little problem of provenance. New information had to have institutional/societal acceptance. The same issues applied to health when the doctor knew all that there was to know and there were no other sources of health information: health knowledge was definitive, if not universally available.

The internet has blown open the floodgates of data, information, knowledge, opinion, rumour and conspiracy theory. Much of human knowledge and information is but a mouse-click away. This glut of information is changing the relationship between educator and educated, doctor and patient. In a world of health-information plenty and no constraints on access, there are huge potential benefits – and, of course, risks. The big questions are:

- How can the benefits be maximized and the risks minimized?
- What will be the characteristics and definable features of health educators and their information resources, and of doctors and their information resources, in the future?
- How will learning change and how can the formal/institutional support reach its potential for converting health learning into healthy action for personal health and well-being in the next decade and beyond?
- What media and channels will be used to provide access to information, knowledge and learning at any time and in any place?

The copyright conundrum

Now there is a growing democracy of 'content' that stacks up the good, the bad and the ugly sources of information and opinion. Internet search does not always deliver the most appropriate results for any given context, although search algorithms are improving and the semantic web offers great promise in this direction. Nevertheless, new and unfamiliar contextual skills are required to sift and filter, question and verify information from the internet. Acquisition of these skills will meet with resistance from a generation educated in the old ways and dangers of knowledge sharing – collaborative working and learning are still being uncovered.

Now the large physical institutions of the past must be replaced by a virtual information infrastructure that guides to information, supports validation of information and supports the application of that information. In addition, resources and attitudes and societal/legal changes are required to ensure that the most valuable ideas of the old epoch are not lost in the mass of competing data/information.

Google is some way towards its goal of digitally archiving all books. The low-lying fruit are those books that are out of copyright, since legal issues bar archiving of in-print books. So out-of-print books are, perversely, the most accessible to the new generation of learners, brought up on instantly accessible, web-delivered knowledge. An article in the *Chronicle of Higher Education* (Barton, 2009) identified the phenomenon whereby old but web-available resources and references are being used preferentially by students in their assignments. Whilst copyright prevents online reproduction of modern text and reference books, the skew towards old, copyright-free works is likely to be accentuated.

Targeting information

A key feature of emerging digital and web technologies in learning is the ability to eliminate the time constraints of accessing information resources in one geographical location at a specific time. Information can now be easily provided at any time and in any place on a wide range of media. Indeed, information can be 'pushed' to target audiences via e-mail, web and mobile phone in ways unimaginable a few years ago. We can remind students where their morning lecture will be; we can remind patients to take their medication or to get their children immunized. We can wake a person each morning and suggest to them their breakfast and indicate the room they must go to for their first meeting or lecture.

Specific strategies for content delivery – and exactly what content is – remain to be determined. However, the technology is there to ensure that the right stuff will be delivered (pushed or pulled) at the right time in the right place. The market/users will get what they want – and perhaps what they need – when they want and or need it.

NHS Choices

The opportunities and the risks are beginning to be addressed in the public-facing website of the UK National Health Service. NHS Choices (www.nhs.uk) is stepping up to the challenge of establishing what medical knowledge services for health might look like to support the public.

NHS Choices aims to arm patients, families and interested others with information about health and the availability of health-related services. It aims to inform choice and to signpost services. It supports choice of healthcare provider (e.g. hospital), choice of treatment (through knowledge about the treatment options) and choice of a healthier lifestyle (using knowledge and tools to inspire and promote healthy behaviour change). Launched in 2007, NHS Choices has rapidly become the most-used health website in the UK, with 8 million visits per month. During peaks of public interest in swine flu in summer 2009, activity reached 12 million visits per month. NHS Choices provides services not only via the web but also by developing and piloting services on mobile devices and kiosks (interactive multimedia stations). These pilots aim to reach people who may be digitally excluded; they use social marketing principles to identify users' needs and best strategies to reach them.

The majority of NHS Choices web content is internally commissioned and links are provided to verified organizations. Increasingly, quality control of these links to external third-party organizations will be driven by an Information Standard certification scheme.[1] NHS Choices provides links for further information to partner organizations such as NHS Evidence, enabling patients and the public to access clinical knowledge sources. It also promotes the concept of partner organizations and provides an open API for partners. Syndicated content is provided free of charge to over 100 distribution partners. For example, NHS Choices has a channel on YouTube, where all NHS Choices videos can be viewed.

Keeping up to date

Keeping more than 60,000 topics up to date can be challenging. NHS Choices is interested in more innovative ways of maintaining and developing content and has embraced user-generated content far more than any government website in the UK. Users can comment (TripAdvisor style) on hospital services and over 15,000 comments have been posted. This facility is being extended to family doctor services in October 2009. NHS Choices users can comment on any page of content on the site. NHS Choices has recently entered into a pilot agreement with Medpedia,[2] a US professional-facing democratic publishing website built on the mediawiki software. NHS Choices syndicates content from the Medpedia user base, who then can access it and update it as they see fit. This updated content could, theoretically, be reintegrated into NHS Choices once it had passed a quality assurance process. This Web 2.0 approach to legacy content offers the possibility of a mechanism for updating by an expert community that could potentially be cheap, quick and of high quality.

Mobile opportunities

With new mobile and web technologies come new opportunities and abilities that may permit library and public health innovators to extend their goals beyond the simple provision of information for learning and application of learning, and to enhance the productivity of the processes of learning and the application of learning. In their influential book *Nudge*, Cass Sunstein and economist Richard Thaler (2008) claim that individuals may be led to new behaviours through careful nudging at opportune moments.

Mobile services have particular characteristics that create real opportunities for access to information and services. Mobiles are:

- always on, frequently carried by users wherever they go: this allows for time-sensitive content to be pushed as required
- location sensitive, enabling providers to serve up content relevant to precise location (e.g. nearest hospital accident department in an emergency)
- ubiquitous, with mobile penetration in the UK exceeding 86% (Ofcom, 2008).

Mobiles also provide the opportunity to engage with an audience that would not necessarily use the internet. Mobile penetration in the young demographic (15–24 age group) is particularly strong. To young people, communicating by text is as natural as by voice, creating a significant opportunity for communication of health messages. Furthermore, while just 49% of unskilled working class and unemployed people have access to the internet, 77% have a mobile phone. Looking at the older demographic, 77% of those aged 65–74 use a mobile phone, while only 41% access the internet.

NHS Choices' mobile provision is less well developed than its internet service, but it is growing, and the potential of this channel is being explored. Mobile services currently provided by NHS Choices include:

- geographically intelligent text-back 'Find My Nearest Service', which identifies local doctors, dentists, pharmacists and a growing list of other services, based on the location of the mobile phone used to text the enquiry
- mobile internet service (www.nhs.uk/mobile) developed in partnership with the UK cross-government digital service Directgov
- mobile apps, including iPhone Drinks Diary
- SMS-based reminder service for HPV (human papilloma virus) immunization
- real-time patient feedback tool to allow patients to give feedback via SMS about their experience of using hospital services
- pilots in using mobile for local health promotion initiatives; these have included SMS motivational fitness messages in the city of Derby and Bluetooth messaging to promote chlamydia screening in the city of Hull.

Digital services

Increasingly, NHS Choices is developing tools aimed at supporting individuals to take control of their health and well-being, not only by providing information but also through more sophisticated interaction. Existing tools include a pregnancy planner, birth-to-5 trackers and a personal health-risk assessor. A more sophisticated approach (requiring sign up and delivery of a much more complex system) is being taken to piloting provision of online cognitive behaviour therapy to encourage mental well-being among the general population. Outbound e-mail alerts are being delivered to registered service users to provide them with timely and specific health information where and when they want it. These first steps have the potential to support individual programmes for health via timely reminders and tools, and to positively influence health-related behaviour.

This area of digitally delivered decision-aid and health-maintenance tools is likely to expand significantly, as it is potentially of great benefit in the area of public health. However, harnessing its potential will require both technological breakthroughs and further shifts in both institutional and public behaviours and expectations in terms of privacy, confidentiality and trust. Equally, identifying what works is a significant challenge and will remain so. Proving that something works and should be adopted in the UK health sector is particularly challenging where comparisons are beginning to be drawn between health technologies and pharmaceuticals, where standards are set by expensive and time-consuming clinical trials. Health technologies are delivering digital programmes that you might find online or on your iPhone as an app (e.g. for losing weight or measuring blood glucose). These can potentially have as much benefit as drugs that are currently offered to have the same function (e.g. reduce blood glucose). So regulators may be beginning to expect the same levels of research and safety regulation around the development of the digital technologies as they do of pharmaceuticals, even though there is a whole realm of biophysiological risk that technologies don't entail but which medicines do. Also, the clear bottom line of the commercial sector does not apply in most health systems, particularly the UK health system, where priorities include clinical issues as well as cost effectiveness, quality and choice.

In the fast-moving world of digital innovation there are neither time nor the resources for large and complex trials. NHS Choices has been developing a middle way for evaluation, whereby the traditional customer-insight, market-research approach of industry has been combined with a more scientific, controlled-trial approach. This has permitted evaluation of workplace kiosks pilots that can offer stronger evidence of the benefit in a language that health commissioners require, whilst minimizing the costs and resource requirements, thus enabling decisions to be made more easily and in a more evidence-based, cost-effective way. A particular example of this approach has been the evaluation of health information kiosks in workplaces in the city of Derby, which have shown that the kiosks can motivate people to make changes for better health.

Conclusion

The old definition of education has been the knowledge and skills acquired through learning. Health education is perhaps the most aspirational of all, aiming not only to educate but to promote and support healthy behaviour change. Knowledge and skills do not necessarily convince the individual to acquire associated behaviours. The potential of push-pull, multi-platform, personalized communication technology in the health sphere is to produce behaviour change. Information technology could go further than educate, and could help health promotion and public health turn a long-awaited corner to realize the potential benefit of social behavioural shift in the population. The challenge to innovators and strategists is clear, the tools may be there and the territory is there to be conquered.

Notes

1   www.theinformationstandard.org
2   www.medpedia.com

References

Barton, T. (2009) Saving Books from Oblivion, Oxford University Press on the Google Books Settlement, *The Chronicle of Higher Education*, 29 June.

Ofcom (2008) The Consumer Experience, www.ofcom.org.uk/research/tce/ce08/research.pdf

Thayler, R. and Cass, R. S. (2008) *Nudge: improving decisions about health, wealth and happiness*, Yale University Press.

15

'Ask us upstairs': bringing roaming reference to the Paley stacks

Fred Rowland and Adam Shambaugh

Introduction

The internet, virtual access to library resources and mobile technologies have all changed settled patterns of library use. In order to find a meaningful role in the new relationship between users, technology, buildings and staff, librarians need patience, persistence and a willingness to experiment with new services. The 'Ask Us Upstairs' initiative, carried out in the fall and spring semesters of 2008–9, was an attempt to bring reference services to the Paley stacks, located on the second and third floors of Paley Library, where most of the Temple University Libraries' book collection is housed. The Paley stacks also include study tables, comfortable seating and computer workstations. 'Ask Us Upstairs' was part of our ongoing efforts to move beyond the reference desk, to other parts of the library and, indeed, the campus, to provide reference services. Though the results were disappointing, many important insights were gained through the careful observation of user behaviour in and around our book collections.

The Samuel L. Paley Library, Temple University's central library, was built in 1966 and comprises four floors – a ground floor (basement) and three floors above. Patrons enter on the first floor, a very large open space with high ceilings, divided into east and west sides. The first floor houses the reference desk and reference collection, over 100 public computers, current journals and newspapers, comfortable seating, and a café. Floor-to-ceiling windows surround three sides, providing a view of important campus thoroughfares. The first-floor entryways to the upper-level Paley

Stacks are fairly well hidden from view and it's often a struggle just to make students aware of them. The lighting upstairs in the stacks is rather dim, as the outer walls were designed to keep out daylight, and the floor plan is uninspiring. The bookshelves are very high and the aisles are narrow, making for a rather unwelcome browsing environment. The infrastructure is outdated, making installation of additional electrical outlets and cable drops complicated and expensive.

The cost to renovate these floors aesthetically and functionally is prohibitive. Despite this, significant improvements have been made, due to a financial boost following the arrival of the new Dean of Libraries in 2005. In addition to building improvements, large ongoing investments have been made in both print and electronic resources. One of the first cash infusions enabled a large purchase of new and back-stock books across the disciplines. The decision seems to have been a good one, as our library's circulation figures have since been increasing steadily. This is consistent with the increase in many of our other public services statistics (see Table 15.1).

Table 15.1 Increasing annual usage statistics, 2005–9

Academic year	Reference desk transactions	Virtual reference	Off-desk reference	Instruct- ional sessions	Paley circulation	Gate count
2005–2006	30,645	1,637	2,849	278	151,674	2,088,207
2006–2007	34,192	2,275	3,037	357	164,329	2,048,051
2007–2008	37,836	1,781	3,750	431	198,184	1,829,207
2008–2009**	38,193	2,223	3,776	657	219,523	2,251,599

** Academic year is July through June. 2008–9 excludes June 2009 statistics.

Users' satisfaction with library services

The results of our 2006 LibQual Survey revealed significant dissatisfaction with library collections and services, including both the physical environment and support in the Paley Stacks. Indeed, our library had been underfunded and understaffed for decades, so the responses were hardly surprising. Providing a multiplicity of study and social spaces in a pleasant environment and finding ways to deliver services efficiently were essential to encouraging students to use the second and third floors. Students have no patience for inconvenience and little sympathy for the dark, mysterious and dusty library shelves of yesteryear. Though the same survey showed strong satisfaction with staff, respondents also found there were too few of us. Students complained of feeling abandoned when trying to retrieve books from shelves. They found the stack areas unappealing and intimidating. There was no formal reference help in the Paley Stacks, though student workers shelving books did answer questions when approached by patrons. Student workers often felt that the questions they were getting amounted to reference questions and really should be handled by reference librarians.

The Paley Stacks were bursting with books at the time of the 2006 LibQual Survey. Not only had shelves been gradually added to the original floor plan, but student shelvers were constantly shifting books around to make space for new volumes, often making the signage outdated. The stacks looked more like a warehouse than a library. Fortunately, we opened an off-site storage facility in 2006, moved hundreds of thousands of volumes there, dismantled rows of shelves and opened up many spaces in our barracks-like configuration. With new funds available, we added comfortable seating and new carpeting in selected areas, computer workstations and many more study tables. This greatly reduced the claustrophobic atmosphere, but the high shelves and the narrow aisles remained. Though we still didn't look like a 21st-century library, the second and third floors were vastly more appealing, comfortable and functional than before. Students responded by using these two floors in increasing numbers for research, study, computer work and, to a lesser extent, socializing. Each of the two floors is divided into east and west sides. On the second floor, east side, group work and group study are welcomed. Most of the students who use the Paley Stacks for study and computer work are undergraduates.

Design of the reference service

In the summer of 2008 a committee of six RIS (reference and instructional services) librarians met and began discussing ways in which we could design a reference service for the Paley Stacks. Most of the physical improvements had been completed and our usage statistics – reference transactions, instructional sessions, circulation and in-house traffic – had been rising annually for a number of years. It seemed like a good time to try to increase our reference presence upstairs. The following are some of the basic decisions we had to make in order to design our service.

- Should we roam, or create temporary reference stations? We decided it would be more efficient to roam, allowing us to cover the whole area and find users right at their point of need.
- How should we make ourselves identifiable in the Paley Stacks? Although some of us saw a need to be highly visible – wearing vests or hats – we decided to wear simple 'Ask Me' buttons (Figure 15.2).
- Which days and times would work the best? Would certain weeks/months during the semester be better than others? After speaking with the stacks supervisor, we decided to provide roaming service Tuesdays, Wednesdays and Thursdays, 10 am–noon and 2 pm–4 pm. We roamed two weeks in December and then the entire spring semester, except for weeks two and three.
- How should we approach users to initiate reference transactions? We came up with a list of questions as prompts: Is this for a paper/project? What kind of paper/project is this for? Tell me more about your project . . . (Figure 15.1).

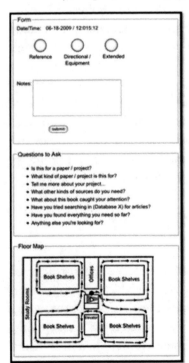

Figure 15.1 Roaming reference form (used to report statistics)

- How should we advertise our new service? We designed laminated posters, which included pictures of each librarian, a map of the route we would take and a description of our new service. These posters were placed in the stairwells and elevator exits and in various high-traffic areas of the second and third floors (Figure 15.2).
- What technology should we use? We had already purchased Apple iTouch hand-held devices for use in the project, though librarians were free to use other devices of their choosing.

Roaming the stacks

Near the end of the fall, we were ready to begin. Starting with the final two weeks of the fall semester and continuing

Figure 15.2 'Ask Us Upstairs' poster, iTouch and 'Ask Me' button

through the spring semester, we made rounds through the Paley Stacks on Tuesdays, Wednesdays and Thursdays from 10 till 12 and from 2 till 4. We started off with a flurry of activity in December 2008. The librarian participants were enthusiastic about the interaction we had with students during the initial two-week period, when students were finishing up term papers and getting ready for final exams. Although the response was not overwhelming, we felt it was encouraging. The spring semester yielded very different results, as activity dropped off a cliff. We began roaming the first week of the spring term, but by the end of that week, due to very little activity, we decided to wait until the fourth week of classes before roaming again. Unfortunately, the activity level did not improve as the semester progressed. Given our experience in December, many of us had expected our shifts to become busier in the last month of the spring semester, but this also didn't happen (Tables 15.2 and 15.3).

Review of results

Clearly, the time that reference librarians put into this project did not pay off in terms of bringing reference service to the Paley Stacks. Twelve hours of roaming for 14 weeks (two in December, 14 in the spring) yielded just

Table 15.2 'Ask Us Upstairs' statistics, December 2008–April 2009

Month	Reference	Directional/equipment	Extended	TOTAL
December	20	29	2	51
January	10	8	0	18
February	12	4	1	17
March	5	4	0	9
April	17	3	3	23
Total	64	48	6	118

Table 15.3 Comparison – 'Ask Us Upstairs' vs other important user statistics

Month	Roving reference	Reference desk transactions	Virtual reference	Off-desk reference	Instructional sessions	Paley circulation	Gate count
December	51	2761	147	277	3	16,654	212,673
January	18	3178	152	297	18	17,136	143,719
February	17	4220	299	494	176	22,736	293,290
March	9	4035	374	361	89	22,859	229,411
April	23	4841	425	250	43	26,910	311,116
Total	118	19035	1397	1679	329	106,295	1,190,209

118 questions, and few of these were complicated enough to require a reference librarian. We frequently receive an equal number of questions in a single day at our main reference desk. There may have been some under-reporting of transactions and there were some missed shifts, but neither of these factors would have significantly affected the results. While 'Ask Us Upstairs' didn't pay off in terms of reference service, it did provide us an opportunity to observe how our patrons use the second and third floors, a sort of informal anthropological study. Several librarians have offices in the Paley Stacks, but few spend much time observing what students are doing, except in the immediate vicinity of their offices.

During our roaming shifts, interactions with patrons were characterized by a certain amount of discomfort. Students were obviously surprised to be offered help and often not certain what to make of it. Typically, they would turn down offers of assistance, but when we did help patrons, it was most often in finding books. To the surprise of some librarians, it was much more difficult to develop in-depth reference transactions from these approaches than we had expected. As one librarian put it, students seemed to have a clear idea of what they intended to do before coming upstairs. They were very directed in their intentions and behaviour –

whether to retrieve books, to work on the computer or to study – and not very open to suggestions from reference librarians. It was rare for students to approach librarians with questions.

Although we had intended to provide assistance in all areas of the second and third floors – the stacks, the group study, computer workstation and comfortable seating areas – librarians rarely offered assistance to patrons in non-stack areas. There was a consensus among librarians that these areas appeared to be 'student-controlled territory' and that student/non-student interactions were not particularly welcome. When students were studying, or working at a computer, it seemed particularly intrusive to offer help. In order to have any sense of whether they needed help, one would need to know what they were working on, and observing them felt a bit awkward, too much like stalking. This perception was pretty consistent among all librarians, regardless of age or gender. Of course, students at study tables and computers could have asked for help, but this they rarely did.

To browse or not to browse

One of the most striking insights we gained was the apparent lack of any serious browsing, despite the fact that our circulation statistics are climbing and our 'in-house' use statistics are as high as our circulation statistics ('in-house' use refers to books removed from the shelves and used by readers on the library premises). Our stacks seem to be mainly used for storage and retrieval, and it was so unusual to see someone browsing that one of us took it as a sign that the patron was lost! As this librarian put it, 'If someone was browsing, I knew he was lost.' Students appear to go to the upper floors to pick up books in the same way that they go to the printer to pick up full-text articles. They browse the library collection through the library catalogue and simply go upstairs to retrieve the items they have found in the catalogue. Whether this is a trend reflecting a change in the way students do research, or is simply the result of our narrow aisles, high shelves and poor lighting is an open question. It could have important implications for the adoption of electronic books. In any case, if we want to encourage more browsing in our collection, we certainly need to move more books off-site and lower and spread out the shelves. If browsing isn't the main priority, space could be used for other purposes, given sufficient availability of off-site storage. There does, however, seem to be one advantage of high shelves: they tend to reduce noise and encourage a quiet study environment, which many students need and appreciate.

Roaming communication

Apple's iPod Touch was used while roaming by all except one librarian, who used a netbook. The iPod Touch turned out to be a poor choice for our project, although it was important for our librarians to begin experimenting with hand-held devices. The iTouch screen was too small to view with patrons and the wireless was unreliable in the stacks, where our most frequent interactions took place. In addition, a number of librarians didn't like the Apple touchpad, one librarian stating that his fingers were 'too thick', another suggesting that the raised touchpad of a Blackberry would have been preferable. On the other hand, the librarian who used a netbook was very satisfied because it was small enough to carry easily but had a large enough screen to view with patrons. She also had no problems with connectivity. There are stand-up computers in various locations on the second and third floors, and many librarians preferred to use these when helping patrons. We did observe that many patrons are now using their mobile phones to track down books in the stacks. In the fall, we enabled a feature in our library catalogue that allows users to text individual bibliographic records to their mobile phones, and students appear to be taking advantage of it.

One of the biggest impediments to effectively providing reference in the stacks was the lack of telephone contact with the reference desk on the first floor. A librarian at the reference desk had no effective way of sending a patron directly to the roaming librarian upstairs. Visibility may also have been a problem. We wore 'Ask Me' buttons and carried our iPod devices to make us recognizable to patrons in the stacks. Our pictures were also on posters throughout the second and third floors. Wearing something distinctive, like vests or hats, might have helped. It is doubtful, however, that either of these two factors would have significantly changed the reception of our new service.

Navigational aids

Most of the questions we received involved helping patrons to find books.

This indicates that we need to improve signage and come up with new ways to help students navigate the stacks. One of our librarians is working on a plan to colour-code shelving signs. Another initiative this same librarian took was to create a guided tour of the library on Flickr, which we can show to users at the reference desk. Yet a third option, which we introduced in March 2009, is a mobile phone tour of the library which allows the user to input the first letter of a call number to find the location of a book in the Paley Stacks. For a few years, we have had wall-mounted phones (along with phone numbers) available in the stacks for calling the various service points in the library for assistance. To our knowledge, however, these have had very little use. Finally, better visibility of student shelvers in the stacks will assist our patrons in finding help. Soon, shelvers will be required to wear vests that will identify them as library employees.

Is virtual reference the answer?

Of particular interest during this period was that our virtual reference interactions increased considerably after we installed a chat widget on our Ask A Librarian web page in late January 2009. It is useful to consider why students responded to this virtual service, while at the same time rejecting our efforts in the Paley Stacks. The chat widget is immediate, anonymous and puts the user in control, and the information usually comes in small chunks. Students are free to multitask as their questions are being addressed, and they can to some extent modulate the speed of interaction. If they feel that they are receiving irrelevant information, they can cut off the chat at any point. Since it's anonymous, they needn't worry about offending the librarian, and they can return to the service at any time. In contrast, when a roaming librarian offers help it's immediate but not anonymous, the user isn't in complete control and information may not be chunked. Certain social conventions come into play so that there are expectations of undivided attention during the interaction, even if the user is not satisfied with the information. Most undergraduates have a hard time breaking off this sort of interaction.

Conclusion

As we noted in the introduction, finding a meaningful role in this new library environment takes patience, persistence and a willingness to experiment with new services. The 'Ask Us Upstairs' initiative was an attempt to fill a need for reference service on the second and third floors of Paley Library. Carefully organized and diligently executed, it failed to meet with even a minimal level of success. Whether the need was more apparent than real, or whether there was simply a mismatch between what we offered and what patrons desired, is difficult to say. However, we learned a great deal about our users and will no doubt continue to seek ways to meet their evolving needs.

16

The role of an agent supplying content on mobile devices

Jose Luis Andrade

Background

Swets is a global subscription services company, building on more than 105 years of experience to maximize the return on investments in time and money for clients and publishers in today's complex information marketplace. As there are many intricacies involved with managing print and electronic subscriptions, Swets offers a portfolio of fully integrated solutions designed to simplify the way information professionals acquire, search, access, manage and evaluate print and electronic resources. Swets is unique, being the sole subscription agent that is ISO (International Organization for Standardization) 9001:2008 certified on a global basis, meaning internal work processes are performed under controlled conditions and subject to continuous improvement.

As is the case in other industries, the expanding body and importance of electronic information suggested that companies needed to find new ways to stay current. With the emergence of the world wide web, traditional print resources were not enough, particularly in an industry that prides itself on delivering the most up-to-date information to its end-users. In a YouTube video published by JISC (Joint Information Systems Committee), *Libraries of the Future* (JISC, 2009), the importance of accessing information anywhere, anytime and any way was stressed. Students want to work wherever they happen to be and to have continuous access to as much content as possible, including through their mobile devices. The rapid increase in the amount of online scholarly content clearly identifies that e-resources have an important role in research, often containing features and elements that may not be able to be represented in traditional print (Luzón, 2007, 59).

To align itself with the value of providing cutting-edge technology to the information community, Swets employs people in the field who seek out trends in the marketplace and sit on library-industry standards committees. In addition, Swets has two customer advisory boards in North America that meet at the ALA (American Library Association) Annual Conference, the ALA Midwinter Meeting and the Special Library Association Conference. Swets received direct feedback from these advisory boards more than a year and a half ago, expressing that delivery of information to mobile devices would emerge as an important trend in accessing electronic content.

The purpose of this overview is to discuss Swets' vision of enabling mobile delivery of content through its suite of electronic products, known as SwetsWise, and to demonstrate how several components allow libraries to acquire, access and manage their electronic content.

First steps

Swets' website and products went through a redesign phase. It was decided that Swets would follow more simplistic design principles, anticipating that users would migrate to mobile devices. As a result, once mobile devices were developed that could fully support HTML, JavaScript and PDF (portable document format), end-users could easily navigate and enjoy the full benefits of SwetsWise. Currently, Swets has fully tested its SwetsWise services on the Apple iPhone and iPod Touch devices. In theory, SwetsWise should operate on any device that fully supports HTML, JavaScript and PDF. The rest of this chapter will describe the types of products Swets offers and how these services are currently used by mobile users, both librarians and end-users (students, researchers) alike.

Acquiring and managing subscriptions

Figure 16.1 shows the SwetsWise interface, as it appears from the Apple iPhone. The capability to search a table of contents as well as full-text articles is one of the many features SwetsWise offers from the respective mobile device.

Figure 16.1 SwetsWise interface

SwetsWise is an integrated solution that can be accessed anytime and from any place. In the current library climate, librarians are pulled in many different directions – to various buildings on campus, to multiple meeting sites or from various conferences nationwide. Information professionals thus need the ability to manage their resources from multiple locations. Through SwetsWise Subscriptions: Library Edition or, for corporate customers, SwetsWise Subscriptions: Corporate Edition, librarians can now manage their electronic resources through their mobile devices. SwetsWise Subscriptions: Library and Corporate Editions are the central focus of SwetsWise, whereby librarians can manage thousands of subscriptions from one seamless web interface. This includes ordering, claiming and renewing subscriptions. In addition, librarians can receive up-to-date information on the status of each subscription, evaluate and report on how users assess the content that the library purchases and control variables that directly affect the needs of demanding users. To streamline an efficient library work environment, librarians can access SwetsWise from a mobile device to research titles, view cost information, find information on how titles can be accessed and whether these titles require a license, and place orders in a matter of seconds. In the current, fast-paced library setting, when every second counts, it is essential to focus on time management.

Accessing content

One of the most important demands Swets has received from the marketplace, especially from its medical customers, is that of having instant access to articles from any place, especially essential peer-reviewed journals, via their mobile devices. Electronic content is an advantage, as end-users can quickly search and receive scholarly information in ways that are not possible in print (Luzón, 2007, 63). For Swets customers using

SwetsWise Online Content, Swets electronic journal gateway, this has become possible. SwetsWise Online Content offers an A to Z list service, enabling the creation of customized lists of all subscriptions, and links to where users can access the content of those subscriptions; offers special features such as integration with document delivery and pay-per-view systems, multi-level-linking technology; covers more than 20 million publications and more than 26.5 million articles; and offers a full overview of and control over electronic journals.

SwetsWise Online Content has been integrated into SwetsWise Subscriptions: Library Edition. When librarians order electronic journal titles through SwetsWise Subscriptions: Library Edition, the orders will be set up in as little as 24 to 48 hours in SwetsWise Online Content. This benefits information professionals, as the work of setting up the access to electronic journals is performed by Swets, and benefits end-users, who receive speedy access to electronic content.

Managing rights

As libraries are ordering more electronic content, it is essential for librarians to be aware of the many terms and conditions of journal licenses. Can the e-journal content be used for interlibrary loan? Is perpetual access guaranteed? Does the publisher support IP access? These are just a few of the questions for which librarians need answers. Managing electronic resource licenses can be complex. In the article 'Managing Electronic Resources' by Marilyn Geller, the point is made that, without an agent, managing your own e-resources could mean getting a current copy of the publisher's license either from the publisher's own website, from a vendor or from a contact within the publisher's organization (Geller, 2006, 8). 'It can be as tedious as making phone calls, leaving messages, waiting for return phone calls, missing phone calls, and finally getting a copy of a license sent to the library for review' (Geller, 2006, 8).

SwetsWise eSource Manager, fully integrated into SwetsWise Subscriptions: Library Edition, solves this tedious process. SwetsWise eSource Manager is a comprehensive electronic licence and resource management solution that currently allows customers to manage licences without manually typing in all the licence details. It delivers a pre-populated database of standard publisher licence conditions that one can easily search, view and customize to reflect specific publisher licence agreements.

Users can customize information in 45 standard publisher licensing and 25 custom user-defined licensing fields. SwetsWise eSource Manager users can view all of this information on their mobile devices and can judge whether the terms and conditions are in agreement with their library's current policies before placing an order for a publication.

Measuring usage and cost of electronic content

The hardships of today's economy bring new challenges to information specialists. As every penny counts, it is imperative to ensure all content to which a library subscribes is relevant and, most importantly, *used*. For many institutions, it is no longer feasible to collect materials solely to keep a complete record of scholarship. Now, more than ever, library directors and deans want to see a return on investment on their materials budget. Cost-per-use has become part of best practice for gauging the viability of electronic resources.

Fully integrated with SwetsWise Subscriptions, SwetsWise Selection Support combines usage statistics with cost information, resulting in a cost-per-use overview for collection analysis. The usage reports adhere to the stringent requirements of the COUNTER Code of Practice,[1] that ensures all usage reports are consistent, credible and compatible. Now, when deciding to renew or cancel a subscription to electronic content or when sitting in a meeting with a director and asked how a particular collection is being used, information professionals can pull up usage statistics and cost-per-use reports at any time from their mobile device, so long as it supports Microsoft Excel viewing. Anywhere, anytime access to usage reports makes collection development analysis more of a science rather than an art.

Searching for all available content – partnerships with technology leaders

Equally important as accessing content and collection analysis is actually finding the information. While search engines such as Google are a sufficient means for finding a good deal of information, there is a wealth of content that libraries own (special collections, internal documents) that can not be accessed through standard internet search engines. In this respect, libraries have come to see the importance of federated searching, a solution that searches multiple sources through one query.

In speaking with members of the library community, it became apparent to us that, especially in the medical and corporate worlds, information professionals were looking for a single product by which to gain access to the majority of their content. As this is a costly technology, Swets went out and searched for a partner who already had tremendous experience in the field of federated searching and whose technology could blend into the SwetsWise interface. The result was SwetsWise Searcher federated search service, powered by MuseGlobal technology. Mobile delivery-compliant, this resource interfaces with a list of more than 4,000 data sources and simultaneously searches through internal and external databases and collections. End-users (students, researchers) now benefit because they can search all of the resources the library owns via their mobile devices and access the content of the resource at any time.

In the internet world, users also want national and international news as soon as it is available. Most libraries can not afford to subscribe to as many individual newspapers as they would like. To solve problems like this, Swets partnered with NewspaperDirect to offer PressDisplay, a web-based portal which provides online access to more than 650 newspapers and magazines from 76 countries in 38 different languages. PressDisplay supports mobile devices such as the BlackBerry and Apple iPhone, so that end-users can have instant access to national and international news. Features included with PressDisplay, such as the ability to share articles through social networks, comment on, listen to, bookmark or customize views of each newspaper make this service ideal for the information community.

Conclusion

By consolidating all of the relevant information in its suite of SwetsWise products, Swets has made the process of ordering, accessing and acquiring electronic information more simple for librarians and information professionals. Further, by wisely designing its website and products so that they fully support HTML, JavaScript and PDF, Swets has enabled mobile access to this information before any of its competitors. Lastly, when Swets partnered with other companies to bring in new technologies, it was important to ensure goals were established with these respective organizations to best adhere to meeting customer requirements.

In this fast-paced information environment, Swets knows and

understands the time constraints of its clients and wants them to have all electronic information at their fingertips – literally. As is clearly pointed out in the JISC YouTube video (JISC, 2009), libraries of the future will go beyond the four library walls. With the emergence of the virtual library, it is important for Swets to ensure that information professionals receive the information they need whenever, wherever and however they want it. 'Online content possesses the ability for end-users to easily search, locate, access timely information, and to control the content' (Luzón, 2007, 70). To support the underlying value of electronic content to library professionals, Swets offers content delivery to mobile devices and will continue to demonstrate its dedication to constantly advancing library technology and focusing on key industry needs.

Note

1   COUNTER Code of Practice,
     www.projectcounter.org

References

Geller, M. (2006) Managing Electronic Resources, *Library Technology Reports*, (March–April), 6–13.

Luzón, M. J. (2007) The Added Value Features of Online Scholarly Journals, *Journal of Technical Writing and Communication*, 37 (1), 59–73.

JISC (2009), *Libraries of the Future*, YouTube, (29 June), www.youtube.com.

17

Portable science: podcasting as an outreach tool for a large academic science and engineering library

Eugene Barsky and Kevin Lindstrom

Background and introduction to podcasting

In January and February 2005, the Pew Internet and American Life Project conducted a survey of iPod/MP3-player users and found that one in five 'age 18 and older' own an iPod or MP3 player (Pew Internet, 2006).

More recently, eMarketer estimated that the total US podcast audience reached 18.5 million in 2007. That audience will increase by 251% to 65 million by 2012. Of those listeners, 25 million will be 'active' users who tune in to podcasts at least once a week (eMarketer, 2008). An Australian academic study that measured undergraduate use and ownership of emerging technologies found that in 2007 more than 70% of undergraduates owned iPod or MP3 players, up from 40% in 2005 (Oliver and Goerke, 2007).

A podcast is defined in the *New Oxford American Dictionary* as 'a digital recording of a radio broadcast or similar program – is typically made available on the web for downloading to a personal audio player'. Podcasting was the 2005 Word of the Year, according to the dictionary editors (McKean, 2005). Podcasts are digital files that can be downloaded and listened to whenever and wherever one wants (Barsky, 2006). Originally, podcasting referred to an audio file that was automatically delivered directly to the listener's device using the XML-based format RSS (Really Simple Syndication) and a feed reader. Rather than the listener having to remember to check for new audio files or tune in to a broadcast on schedule, the feed-reader software would automatically check for and download any new audio to the listener's device. Recently, podcasting has become synonymous with any audio or video file that listeners download and play on a digital player (Worcester and Barker, 2006).

Podcasts in university education

The portable audio device is no longer simply a medium for music or video entertainment; it now conveys a lot of educational material. Podcasting usage in education is increasing. With the potential to change the teaching and learning experience significantly, it can facilitate organization and delivery of information tailored to users' individual preferences and learning styles (Harris and Park, 2008).

Podcasts are asynchronous and allow for infinite review and reinforcement of the skills presented. Long files can be broken into smaller, more digestible chunks than typical instructional sessions in academia (Griffey, 2007). The flexibility and affordability of podcasting cater to diverse student needs by enabling repeated review for learning and providing for the effective use of time. Podcasting is also a communication enabler, reaching out to a wider community. For instance, podcasting provides lifelong education opportunities for alumni and creates a culture of knowledge sharing and interdisciplinary collaboration (Harris and Park, 2008).

Podcasting allows existing educational audio content to be made more widely available to various communities and enables educators to develop custom audio content. It provides educational benefits to students who have a preference for auditory learning and for those with sight and/or auditory impairment who rely on audio technology, and it may greatly assist non-native-tongue speakers, since they can review the content many times (Palmer and Hall, 2008).

The popularity of podcasting has grown in academia. In a well-publicized move, Duke University gave iPods to its entire 2004 freshman class as part of a university initiative to encourage creative uses of technology. Students used iPods to listen to podcasts of class lectures and music, to store and transfer files, to record interviews and create their own podcasts (Worcester and Barker, 2006). The Vanderbilt Center for Science Outreach created Snacks 4 the Brain, a podcasting feed that connects working scientists with students and teachers in K-12 classrooms worldwide. Moreover, Stanford University, MIT and other prominent institutions offer lectures and other content to the public free of charge, via the Apple iTunes U online website available (Lee et al., 2008).

Science podcasts at the University of British Columbia

The University of British Columbia is one of the largest research universities in Canada and has a strong Faculty of Science. During the course of an academic year the Library makes hundreds of talks available for the students, faculty and the broader community to access. As librarians, we felt that most of this information is lost quickly after a talk is completed; it could be very beneficial if we could capture those lectures in a digital format and share them with the world.

As a result, the librarians offered to help the Department of Physics with creating, hosting and maintaining podcasts. Our offer was enthusiastically supported by faculty and students and we have recorded a number of physics lectures directed at students, faculty members and the general public. We were successful enough to record two Nobel laureates' lectures as podcasts. We host our podcasts on cIRcle, UBC's Institutional Repository,[1] and provide links from the UBC library website. This initiative was very well received by our community and podcasts were downloaded thousands of times in the first few months. Below, we share our experiences and insights into creating podcasts for the sciences.

Podcasting 101: a step-by-step guide for the sciences

1 First, find appropriate content. Content is crucial – substance trumps style. Quality will keep people coming back for more. We usually work with departments to select workshops and lectures for recording. Copyright is a major issue. We have a simple copyright waiver form for presenters so that we can deposit our podcasts online and make them publicly available.

2 Gather required hardware and software. Podcasting is very simple and cheap. We use free open source software called Audacity[2] to record our podcasts and convert files into MP3s. We use a USB microphone plugged to a laptop to capture sound. Those microphones are reasonably priced and can be found in any music store for less than $75. We use the University of British Columbia's Institutional Repository, cIRcle,[1] to store, archive and maintain our digital podcasts. Many universities in the developed world have institutional repositories in place to capture and archive university-generated knowledge. Alternatively, a departmental website can be used or podcasts could be stored freely on the web using sites such as www.podbean.com.

3 After the final audio file has been converted to MP3 format and uploaded, it needs to be streamed using an RSS feed. This is what turns an audio file into a podcast – when the listener's feed reader software (such as Bloglines[3] or Google Reader[4]) automatically checks and downloads any new audio to the listener's computer/device. One simple way is to create a blog with an embedded RSS feed and point to a podcast as a blog entry. Listeners can then subscribe to the blog's RSS feed and the podcasts are downloaded to their audio devices or computers. Of course, audio files can also be linked from a website for manual download, and in many cases today this is also referred to as a podcast.

4 Promote the podcasts. We advertise within our university community and encourage faculty, students and staff to subscribe. Students can be the best word-of-mouth promoters, but the quality and timeliness of the podcasts will be what keep subscribers coming back for more.

5 Lastly, evaluate and learn from mistakes. We believe that this is one of the most important steps in the whole process.

Conclusion

Podcasting is another method that academics can use to reach young user groups and maintain relevance in their university learning experiences. The versatility of this technology may increase student satisfaction and instructional flexibility. Most importantly, the use of iPods or other portable media devices has the potential to integrate formal education with other aspects of the student's life such as communication, entertainment and work.

It is difficult to predict the outcome of integration, but then this has never before been possible on such a large scale. It is likely that podcasting will bring unexpected and even disruptive aspects to traditional pedagogy and educational processes (Harris and Krousgrill, 2008; Ralph and Olsen, 2007). We have certainly found our podcasting experiences to be mutually beneficial for the library and our academic community. They have fostered relationships between students and their respective faculties.

Even though podcasting is at an early stage, we expect it to continue to grow in popularity as it receives more mainstream press and as new tools arise (for both content creation and delivery to the end user). In addition, increased awareness of what iPods or other MP3 players can hold – not just MP3 files but other types of content, such as pictures, video and text – should enhance their profile. The podcasting phenomenon will grow in ways that we haven't even envisioned as yet.

Acknowledgements

The authors would like to acknowledge Dean Giustini and Glenn Drexhage for reading and commenting on the preliminary version of the article.

Notes

1   https://circle.ubc.ca.
2   http://audacity.sourceforge.net.
3   www.bloglines.com.
4   www.google.com/reader.

References

Barsky, E. (2006) Introducing Web 2.0: weblogs and podcasting for health librarians, *Journal of the Canadian Health Libraries Association (JCHLA)*, **27** (2), 33–4.

eMarketer (2008) *Heard the Latest About Podcasting?*, www.emarketer.com/Article.aspx?id=1005869&src=article1_newsltr [accessed 27 July 2008].

Griffey, J. (2007) Podcast 1-2-3, *Library Journal*, **132** (11), 32–4.

Harris, D. A. and Krousgrill, C. (2008) Distance Education: new technologies and new directions, *Proceedings of the IEEE*, **96** (6), 917–30.

Harris, H. and Park, S. (2008) Educational Usages of Podcasting, *British Journal of Educational Technology*, **39** (3), 548–51.

Lee, M. J. W., McLoughlin, C. and Chan, A. (2008) Talk the Talk: learner-generated podcasts as catalysts for knowledge creation, *British Journal of Educational Technology*, **39** (3), 501–21.

McKean, E. (2005) *The New Oxford American Dictionary* (2nd edn), Oxford University Press.

Oliver, B. and Goerke, V. (2007) Australian Undergraduates' Use and Ownership of Emerging Technologies: implications and opportunities for creating engaging learning experiences for the net generation, *Australasian Journal of Educational Technology*, 23 (2), 171–86.

Palmer, S. and Hall, W. (2008) Application of Podcasting in Online Engineering Education. *International Journal of Engineering Education*, 24 (1), 101–6.

Pew Internet (2006) *Podcast Downloading*, www.pewinternet.org/PPF/r/193/report_display.asp [accessed 31 July 2008].

Ralph, J. and Olsen, S. (2007) Podcasting as an Educational Building Block in Academic Libraries, *Australian Academic & Research Libraries*, 38 (4), 270–9.

Worcester, L. and Barker, E. (2006) Podcasting: exploring the possibilities for academic libraries, *College & Undergraduate Libraries*, 13 (3), 87–91.

Part 4

M-libraries and learning

18

Mobilizing the development of information skills for students on the move and for the workplace: two case studies of mobile delivery in practice

Hassan Sheikh and Anne Hewling

Introduction

Increasingly, students come to their studies equipped with access to a number of mobile technologies. These may include laptops and mobile internet and – even more frequently – access to the internet by means of web-enabled mobile phones. There is much debate about the need for flexible content delivery to meet the perceived needs and expectations of students using such tools. What might constitute appropriate content? Can such devices deal with 'new' content or are they better harnessed for revision and reinforcement activity, as a supplement to other sorts of delivery rather than as stand-alone material? Or, does mobile delivery offer an entirely new kind of study and learning experience and, if so, when and where in the curriculum might this be appropriate? There is also, as yet, little literature available on the actual implementation of such mobile initiatives, the nitty-gritty, for example, of going beyond shrinking existing web pages to fit a mobile phone screen and writing content in a format that is tailored for the smaller screen. This chapter will look at case studies of two real-world projects (mobileSafari and iKnow) which have produced a range of information-skills mobile content that is intended to be delivered to learners in both formal and informal learning contexts, the one as part of a formal higher education programme and the other as part of a workplace-based initiative. Instructional design issues (including the creation of pedagogically sound generic templates), will be considered alongside related technical issues.

The mobile delivery in practice: real-world examples
Safari and iKnow

Safari[1] is an online, open-access course designed to improve information literacy skills amongst Open University (OU), UK students. The course was conceived and developed and is maintained by the OU Library. In 2007 it was six years old and in need of rebranding and revision to include online skills, particularly techniques for evaluating web content. During the review we considered how and when students used the web resource. Consequently, in addition to a web version, we decided to develop a mobile device edition for access via phone networks, which might achieve better and wider access beyond that achieved by the existing web version.

iKnow[2] is all about information and knowledge at work, and offers a variety of free resources aimed at helping users to find and manage information effectively at work. The activities have been created in bite-sized chunks, so that users can fit them in whenever they have a few free minutes, and they can be done either at their desk or on the move.

Adaptation for mobile devices

Although not excessively text heavy, the original content for Safari was not immediately suitable for mobile delivery. The ADR (auto detect and reformat) technology could render pages for mobile delivery but content was not optimized by this means, since the process simply reduced page size rather than remodelling the actual content. We reworked the Safari content into a series of small learning objects that were based on the principle themes from the original content. These were structured using ADR software developed by Athabasca University (Canada) for a screen restricted to around 150 characters or a single image with minimal scroll, only sufficient to display responses to in-content questions. A couple of ready-to-use learning object examples are shown in Figure 18.1.

Content model for smaller screens

As part of our work on mobile services development, particularly around information literacy, we have been experimenting with learning objects for smaller screens and have designed a content model for instructional designers to author mini learning objects that can have a maximum of 5 pages, with no more than 90 words per page. These mini learning objects render

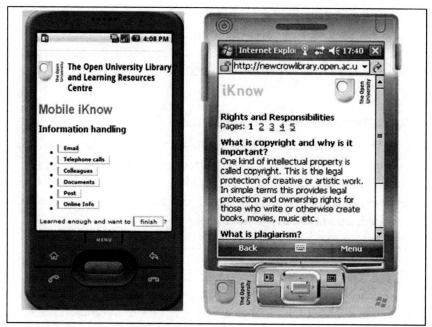

Figure 18.1 Learning object examples

quite well on smaller screens, especially on lower-end mobile phones. This content model has been delivered through our in-house developed Learning Object Generator (LOG) system, which is a web-based application. Using the LOG system, either authors can preview the authored learning objects as web pages, or the system packages the content of those learning objects according to the IMS content package standard[3] for deployment directly into Moodle (the open source VLE used by the OU) and to render nicely on mobile phone screens. These mobile-friendly learning objects have been reused for both Safari and iKnow websites. We have recently evaluated the LOG tool as part of one of the Second International M-Libraries Conference workshops. The participants (in groups of four or five) successfully produced several mobile-friendly learning objects in their chosen information literacy topics. Examples of this work are available on the OU Library's website.[4]

The online home for mobile learning objects

The first set of modules produced for the Safari programme is available for evaluation via the OU Library's website (http://digilab.open.ac.uk/testarea/mobileSafari/index.php).[5] In addition to being used as a stand-alone resource the materials are also available for embedding into course materials in, for example, a VLE, or into online assessment guides, where they serve as revision aids. We have also produced some further mobile-friendly learning objects as part of the recent overhaul of the iKnow website.[6]

Collaboration with Athabasca University Library

Currently, in partnership with the Athabasca University Library team, we are working to further enhance the ADR software. It is now being tested on different devices and platforms, including Nokia N95, iPhone, Qtek, Blackberry, Symbian OS, OS X (Safari web browser) and Windows CE before final modifications are implemented.

Conclusion

Many hurdles have been overcome but there are still fundamental issues that remain at least partially unresolved. The lack of a common platform for mobile devices makes it difficult to optimize content for every device that students may possibly wish to use. Most mobile platforms don't support advanced html features, and this constrains content. Most current mobile internet tariffs are relatively expensive and with limited speed; this means that most users only access text-based content or download small chunks of information. What can be displayed is likewise limited by the small screen size of most devices. Equally, all these factors present a challenge in terms of the pedagogy of mobile learning design.

Notes

1   Safari,
      www.open.ac.uk/safari/.
2   iKnow,
      http://iknow.open.ac.uk.
3   IMS Content packaging specification,
      www.imsglobal.org/content/packaging/.
      For more information about IMS Global Learning Consortium see
      www.imsglobal.org/content/packaging /index.html.
4   http://library.open.ac.uk/digilab/testarea/mLib2009Outputs/.
5   http://digilab.open.ac.uk/testarea/mobileSafari/index.php.
6   http://iknow.open.ac.uk/mobile/.

19

The library's place in a mobile space

Graham McCarthy and Sally Wilson

Introduction

This chapter describes one university library's exploration and implementation of mobile services and its subsequent contributions to broader, campus-wide mobile services initiatives.

Ryerson University is a mid-sized urban university with a diverse student population of 23,000 undergraduate and 1,950 graduate students enrolled in five faculties: the Faculty of Arts; the Faculty of Communication and Design; the Faculty of Community Services; the Faculty of Engineering, Architecture and Science; and the Ted Rogers School of Management. Ryerson also has more than 65,000 annual registrations in its G. Raymond Chang School of Continuing Education, which is the largest in Canada. The university is primarily a commuter campus, with the majority of the students using public transportation. The library has a staff of over 85, including 27 librarians with a strong service focus. The collection consists of over 500,000 books, 2,500 print serials and an extensive electronic collection of over 35,000 journals, 90,000 e-books and a wide array of online databases.

Environmental scan

Prior to fall 2008, the only true mobile service offered by the library was a text-messaging service from the catalogue that enabled users to text the title, location and call number of an item to their mobile phones. This information could then be viewed when the user was in the stacks, looking for the book. The moderate successes of this service and the subjective impression that students were increasingly using smart phones encouraged the library to look more closely at the mobile environment as a venue for its services.

To confirm our sense that smart phones were becoming more popular, particularly amongst the student population, we looked at several surveys conducted in 2008. The Pew Internet and American Life Project's March 2008 report, *Mobile Access to Data and Information*, indicated that within the 18- to 29-year-old demographic 85% were using text messaging and 31% were accessing the internet on their mobile devices (Horrigan, 2008). And in July 2008 a Nielsen report stated that 15.6% of US mobile phone subscribers used the mobile internet (Covey, 2008). Because neither of these reports commented specifically on the Canadian situation we had some concerns that the higher cost of data plans in Canada and the late introduction of the iPhone (in July 2008) might result in a different adoption rate in our environment.

Survey methodology

To get a better idea of mobile device usage amongst our students, we conducted a 13-question survey that ran from 3 to 14 November 2008. The survey was available via the library's website and in print at the two main service points in the library – the circulation desk and the reference desk. The main objectives were to find out what types of mobile phones were in use, what the devices were being used for, what services people wanted on their next phone, when they upgraded, and what library services our users were interested in seeing. A draw for one of two iPod Nanos was used as an incentive to participate in the survey. This resulted in the completion of over 800 surveys, 84% of them by undergraduate students.

Findings

Results from the survey[1] were quite similar to those reported by Pew. Approximately 20% of our users indicated that they had access to the internet on their phones, 85% were text messaging, 63% taking photos, 38% listening to music and 20% accessing e-mail. A key question for us, because we wanted to plan for the future, was 'What will your next phone be?' Users' responses indicated that 38% would get a smart phone and 37% were undecided. This meant that within the coming three years (most mobile phone users in Canada have three-year phone plans) a sizeable number of our users would have access to a smart phone. Our anecdotal evidence of the increased prevalence of smart phones, coupled with these survey results, provided a clear enough indication that it would be advantageous for the library to provide more services to mobile devices.

Service developments

To help us determine where to focus our mobile development efforts, we asked the question, 'What library services would you like to be able to access on a cell phone?' On the basis of the responses to this question we decided to implement those services most desired by students that meshed best with the limitations and advantages of smart phones (ease of use, convenience, fulfilling an immediate need) and with the library's capacity for development. The most highly ranked service requested was the ability to book a group-study room. This was not surprising as group-study room space is at a premium on campus; booking these rooms via the library's website is very popular (17,000 bookings in the fall 2008 semester).

The second most-requested service was 'checking hours and schedules'. Our intention for this option was library hours and library schedules for research workshops, but it became apparent from the high ranking and from subsequent student focus groups that students wanted access to their own schedules. While we were not able to implement this service in our initial mobile offerings (as student schedules are not managed by the library), we were able to provide it at a later date, when the Ryerson mobile project launched in the fall of 2009. Preliminary usage statistics for Ryerson Mobile[2] from early September 2009 indicate that the student schedule service is by far the most heavily used of the seven services currently offered.

The next two most popular library services requested were the ability to check the catalogue and to check the borrower's library record for fines, items charged out and holds. All these services were implemented in the winter semester of 2009. We also added two other services – Find a Computer and Check Laptop availability. Although not specifically requested by students, they were implemented because the nature of these services – quick look-up of real-time information – was ideal for a mobile service. Statistics are not available for all services, but statistics for study room booking indicate that 5% of bookings are currently performed from mobile devices.

At the same time as the library was developing mobile applications for its services, various other groups on campus were developing and exploring their own mobile initiatives. Students in Free Enterprise (SIFE Ryerson) was conducting focus groups to determine what mobile services students thought would improve their campus experience. The Ubiquitous and Pervasive Computing Lab was exploring location-aware services for iPhones and Google Android phones, and the president of the university was promising students access to their schedules on mobile devices. In late winter a project was launched, tentatively called Campus Assistant, to further campus-wide mobile development. The development began in May, when the library hired two students and Computing and Communication Services (CCS) hired one student to repurpose existing services and create new applications capitalizing on the functionality of mobile devices. The major objective of the project was to have students drive the development and design of services, so it was essential to have student involvement from the beginning of the web application's life cycle.

Ryerson mobile

From the analysis of survey results, focus group findings and exploration of mobile projects that other academic institutions had already created, we were able to clearly identify the mobile services that would meet the needs of our campus community. At the M-Libraries 2009 conference (23–24 June 2009) we unveiled an early build of the mobile web applications (previously called the Campus Assistant) and received much positive feedback from the attendees, as well as some recommendations for feature improvements.

The service was launched to our campus community (under a new brand, Ryerson Mobile) on 14 September 2009, as part of the Inaugural Mobile Innovation Week then running in Toronto. Essentially, Ryerson Mobile is a suite of web applications designed to make one's campus experience easier and more enjoyable. For the launch we created seven applications, one configuration profile and an FAQ/About page. A brief description of each application and the configuration profile is provided below.

Campus directory: Provides access to the faculty and staff directory. Hypertext links for phone numbers and e-mail addresses initiate phone calls and e-mail messages if supported by the device, and office locations are displayed on the mobile mapping application.

Class schedule: Lets students view class information for the current and following semester. Schedules can be viewed by daily or weekly time periods or by course code.

Campus map: Shows the location of Ryerson buildings, based on selections from a drop-down menu. This action places a push-pin on the desired location and indicates the street address.

Book a study room: Allows students to search for and reserve study rooms in the library.

Library: Provides patrons with the ability to search the mobile catalogue, place holds, renew books and check fines.

Find a computer: Shows real-time status of computers available in the library and across campus. It also provides availability information for the laptops that students can borrow from the library.

News: Allows one to see the top news stories coming from various sources on campus, including the library's own news feed.

Profile configuration settings: Provides personalization options for users to customize their experience.

The library's role in the project

The majority of the applications developed for Ryerson Mobile do not have the library as the primary focus, yet the library had a huge hand in the development of all of these services. Many factors contributed to putting the library into such a leading development position.

The most important factor was that the library had already experimented in mobile development by setting up a mobile version of the catalogue (AirPAC) and by developing a mobile study-room reservation application. The library also had a desire to provide students more access to services on their mobile devices. Other factors included the persistence of the library's technical team to develop innovative services, an excellent relationship with the campus's central computing (CCS) and the library's desire to branch out and form relationships with various departments and faculties. All of these factors created a perfect atmosphere for development of the campus-wide mobile initiative.

The library played a big role in the development of the web applications, but many different hands in many capacities brought this project to reality. The CCS provided access to databases and APIs for accessing computer availability, staff directories, campus holidays and class schedules. To supplement the findings of the library's survey, SIFE Ryerson conducted focus groups with students from various faculties early in the project's development. SIFE Ryerson also led the marketing initiatives by creating promotional materials and launching a very successful advertising campaign.

Technical implementation

The design is a three-tiered client/server architecture running Windows Server connecting to various web services. These services transfer information to an Oracle database management system that manages our user data. The user interface for the mobile application has been customized for each supported mobile device (e.g. iPhone/iPod Touch, Google Android, BlackBerry 8000+, Samsung and Nokia N Series). The user interface experience on each device is a little different because of the way that the page is rendered by the devices' browsers. We detect devices based on the user agent specified in the browser. Since our mobile application is not native to any specific device's framework, it was important to create a platform diverse enough to cater to a wide array of display specifications.

Future of the project

This is only the beginning of a campus-wide mobile initiative. The seven web applications are just a starting point for the kinds of services that we want our campus community to have access to at any time, anywhere. We are now creating web services to our data and an extensible framework that will enable students to create their own applications that can be contributed to Ryerson Mobile, benefitting the greater campus community. The construction of an environment where collaboration and innovation are highly encouraged will enable the students (through repurposing data and creating mash-ups) to create truly imaginative applications that can directly benefit their peers. This is how we envision the future of the project.

Pitfalls and emerging issues

The first barrier that we will have to overcome in the future expansion of the project is maintaining the highest level of protection of the students' university account information. Opening up a framework that allows third-party developers raises the issues of protection of account passwords and privacy of information. We are currently looking at the Facebook F8 platform and Apple's App Store approaches to third-party application development in order to figure out best-practice solutions.

The second issue that we currently and will continually have to deal with is supporting a wide range of mobile devices without having the ability to test each one properly. The wide variation in screen resolutions, browser configurations, scripting capabilities and bandwidth allocations makes it almost impossible to create a truly personalized and streamlined experience for all of the devices. What we have done instead is to cater to the devices with the widest market appeal. As we continue to promote the service throughout the campus we are asking for feedback about any problems or display issues that are occurring, in the hopes of solving each device user interface issue on a case by case basis.

Conclusion

The library's profile as more than just a curator of collections and space was significantly enhanced and valuable experience was gained during the initial round of implementing and developing mobile applications. When discussions on the broader, campus-wide mobile initiative started in the spring of 2009, the library was well positioned to play a leading role in providing services to the mobile environments frequented by students. The library has traditionally been a steward of information, with a strong service-oriented ethic; however, it has not always had the technical capacity to create truly innovative services. Forming relationships with partners on campus has enabled the library to meet its service goals and to contribute its strengths to broader campus initiative such as Ryerson Mobile.

Notes

1   Ryerson Library Mobile Survey summary,
    www.ryerson.ca/library/msurvey/

2   Ryerson Mobile,
     www.ryerson.ca/rmobile

References

Covey, N. (2008) *Critical Mass: the worldwide state of the mobile web*, Nielsen Mobile,
    www.nielsenmobile.com/documents/CriticalMass.pdf
Horrigan, J. (2008) *Mobile Access to Data and Information*, Pew Internet and American
    Life Project, (March 5)
    www.pewinternet.org/Reports/2008/Mobile-Access-to-Data-and-
    Information.aspx

20

M-libraries in distance education: a proposed model for IGNOU

Seema Chandhok and Parveen Babbar

Introduction

The mobile revolution has swept around the world. During the last decade, due to enormous changes in the mobile industry, mobile usage has become a necessity for the common person. The introduction of a range of advanced features and convergence of all forms of media, such as internet, radio, television, etc., have increased the potential of mobile devices, which can now perform a variety of functions in every sphere of life.

India is the second-largest mobile market in the world and it is expected that the mobile base will expand to 500 million subscribers by the year 2010. It currently has 320 million potential mobile subscribers, with an average of 10 million new subscribers per month (Siliconindia News Bureau, 2008).

Mobile technologies have made a real-time contribution not only to human life generally, but also in the teaching and learning experience, making their value both self-evident and unavoidable. Mobile learning methods offer valuable possibilities for learners in remote and distant parts of the world. Distance education institutions in India are now in the process of adopting mobile electronic equipment and communications technologies to bring the new phenomenon of mobility to distance learners. Indira Gandhi National Open University (IGNOU) has the motto of 'Education anywhere and anytime' and, to achieve this, the university has introduced the concept of 'm-education' and is developing mobile compatible-content for students.

M-learning in distance education

Distance education in its present form has undergone a transition from print, to e-learning, to m-learning. E-learning was the stage of moving towards total automation in teaching and learning processes, using Learning Management Systems (LMS). E-learning has facilitated the development of programmes that utilized internet-based technologies for colleges, universities and research institutions embracing both open source and proprietary LMS tools. E-learning models are already functional in various open universities such as the UK Open University, Hong Kong Open University, Athabasca Open University, Indira Gandhi National Open University, etc. With the wide use of mobile technology, m-learning is becoming an extension of e-learning, and has the potential to make learning more widely accessible to distance learners than was previously possible in distance learning environments. M-learning is a rapidly growing development and has the advantage of allowing learners to learn 'on the move'. As technology has become more widely available, educational options are continuing to expand. For learners who need flexibility, owing to constraints of family, work and mobility commitments, m-learning may be an appropriate option to consider.

The relevance of m-learning to distance education has proved its significance in the current climate. Since mobile devices are now within the reach of most learners, any information that is provided on mobile devices is widely accessible to learners. The course materials and video lectures offered by distance education institutions can be downloaded by learners onto their mobile devices and the same can be accessed whenever there is a requirement. Most of the elements of distance education can be provided via mobile devices. However, the learner's mobile device needs to be enabled for compatible technology (Suresh, 2008).

Benefits of the mobile web for distance learners

Distance education with m-learning is truly coming of age and offers huge scope for innovative solutions. It has the strategic potential to address problems of access to higher education in developing countries, and it supports organizational training and professional development in various new ways. M-learning offers a design model for distance learning that is enhanced by the potential for learning anywhere and anytime; global interaction with fellow students; access to formal and informal learning environments; and engagement in the process of creating learning resources.

Today, distance learners have the opportunity to access the world wide web via a range of mobile devices, from a mobile phone to an iPod, or pocket PC, etc. Mobile web has capability for searching and browsing the internet from anywhere via its mobile signals. Distance open universities around the world are creating m-websites with scaled-down versions of desktop screens, comprising numbered menu systems for quick access to content. Web destinations that do not have mobile versions appear as if they are being squeezed onto the tiny screen, and often have overlapping menus and links. If accessed by way of a search engine, a website may be 'transcoded' or have some formatting applied to it so as to make it more viewable on a phone.

The mobile web is the internet for the small screen, and provides many of the same benefits as its desktop counterpart, such as:

Information access, wherever and whenever: Web-enabled mobile devices provide owners with round-the-clock access to the internet, regardless of location. By freeing information from the restrictions of a desk or search for a nearby Wi-Fi hotspot, they enable distance learners to quickly retrieve and exchange information. Accessing the internet on a mobile device is all about getting the content that the distance learners want, when they want it. Local information on the mobile web, such as counselling sessions, assignments, exams schedules, and so on, satisfies learners' immediate needs.

Limitless access: As mentioned above, the mobile web is not restricted to those sites that have been specially designed for mobile browsing, but encompasses the whole web. Web-enabled phone users have access to all the same online resources that they would find via their desktop computer.

Mobile data potential: Learners can download the information to their pocket PC or smart phone and can move about with the data, reading it at their convenience, even after switching off the internet.

Interactive capabilities: The mobile web offers users the participatory experience of the read/write web in the palm of the hand. Users can create content such as assignments or project reports, upload videos taken with their camera phones, share and receive exam marks and comments, write blog posts, tag resources and form connections on educational networks.

Location-aware: Many of today's smart phones and pocket PCs have global positioning system (GPS) capabilities, making them aware of where they are at all times. Distance learners can search for study points near their locations, retrieve directions to a desired destination and discover nearby peer groups and contacts. (Kroski, 2008, 6–7)

IGNOU at a glance

Indira Gandhi National Open University is the largest university in the world, with 2 million enrolled students taking 15% of total higher education courses. It serves the educational aspirants through 21 Schools of Studies and an elaborate network of 62 Regional Centres, 2300 Student Support Centres in India and 52 partner institutions in 33 countries. The university offers more than 1600 academic courses under 175 programmes of study (IGNOU, 2009a). It acts as a National Resource Centre and functions as a leading apex body to promote, co-ordinate and maintain standards of distance education around the world.

Mobile education in IGNOU

Mobile education or m-education is a new model in the history of IGNOU, using mobile and wireless technologies for its distance education system. Aimed at fulfilling India's 11th Five Year Plan motto, 'Education for All', it aims to take education to the marginalized and disadvantaged members of society. IGNOU is encouraging peer collaboration via wireless mobile devices, PDAs and hand-held devices so as to create opportunities for discovery and education among the distance education community. M-education in IGNOU is expected to make use of collaborative activities in synchronous and asynchronous modes with its learners and their peers. Through m-learning, distance learners will be provided wireless virtual classrooms on their mobile devices in order to support the learning activities of teachers, students and peers in a distributed environment.

In collaboration with Communication and Manufacturing Association of India (CMAI) IGNOU and other players are contemplating providing educational content to students via text, audio and video on mobile phones. Even the ability to take exams on a mobile is a future possibility. It is estimated that the mobile education campaign of IGNOU will cater for 20 million students in 2009, and twice that number in the following year. However, in India Ultra Low Cost Handsets (ULCH) are mostly used, which raises the difficulty of content delivery. While text and audio content can be easily delivered to most phones, distributing video content would require a student to have a video-compatible phone (IGNOU, 2009b). Furthermore, 95% of IGNOU course material, including about 1600 videos, is already accessible through an institutional digital depository e-Gyankosh. Hence, e-learning has moved from being merely a content repository and emulating classroom teaching to the more dynamic concepts of social networking, do-it-yourself (DIY), the personal learning environment (PLE) and m-learning (Kanjilal, 2008).

Learners will soon be able to access IGNOU's content on mobiles and over broadband, as the university has a tie-up with city-based Time Broadband, which is launching Internet Protocol Television (IPTV) services in India. To this end, a memorandum of understanding has been signed between Time Broadband Services Private Limited (TBSPL) and IGNOU. This is a think-tank instituted for the first time in the world of university education systems for the convergence of telecommunication and entertainment. It is expected that two IGNOU channels, Gyan Darshan I and II, will be carried by the TBSPL cable network service round the clock and all through the year. The TBSPL will have the right to downlink signals from the channels. The tie-up has ensured that IGNOU TV will bring an abundance of multiple global digital and audio-visual content to its IPTV consumers via broadband. This will also be available to learners both in India and in South-East Asian partner institutions. The Gyan Darshan channels are reaching out to the masses via the DTH (Direct-To-Home) platform and webcasting. The webcasting of these channels, which are also mobile compatible, has already been tested and verified. The university also plans to offer capsule courses on dance, music, art, public health and other subjects via mobile phone, and these will be available in the near future (IGNOU, 2008, Foundation Day Special).

Mobile SMS service of IGNOU

Two-way SMS service is becoming very popular as a means of requesting and downloading data to the mobile phones. In this instance, the distance learner is required to register a mobile phone number and verify it by using the code sent via SMS. Once registered, the user can subscribe to various SMS channels. Alerts can be sent via SMS, just as e-groups do through e-mails. IGNOU wants to harness the power of SMS to provide student support which is time consuming to do through other modes. An SMS-based information system will provide students with information ranging from their enrolment status to results, and help students and the institution to interact. SMS also has a broader aspect, as it converges with IPTV to allow learners to interact with their coordinators almost instantly. This move further strengthens IGNOU's Open and Distance Learning (ODL) capabilities, in which a conventional system of face-to-face education delivery is converged with appropriate ODL technologies.

Future SMS-based services will include dissemination of end-of-term and entrance examination results; facilities for students to check their marks and to retrieve information from their profiles; updating of learner profiles; etc. This will permit a better student support system, with timely dispatch of course material, information for learners about exam schedules and posting of latest news and other updates.

M-initiatives @ Library and Documentation Services

The Library and Documentation Division (L&DD) of IGNOU has a goal to develop appropriate collections in various disciplines to meet the needs of the diverse users of the libraries at the headquarters, regional and study centres. L&DD at IGNOU is making every effort to take higher education to those whom it has hitherto not reached, through its various information and document delivery services. The most recent initiative is the formation of a Consortium of Open and Distance Learning Institutions, the National Open and Distance Learners' Library and Information Network (NODLINET). This will provide a platform for libraries and information centres of India's ODL system to provide its users with access to electronic and digital resources from leading publishers and vendors around the globe – from anywhere, at any time, using advanced technologies to provide education on a par with that of the conventional educational system (Arora, 2007).

With a growing number of publications, limited resources and a scattered clientele in ODL system, mobile technology is indispensable in meeting the ever-increasing requirements of users. At present, access to

NetLibrary has been provided to Regional Centres of IGNOU around the country via mobile phone. More databases, e-resources and digital texts will be made available to all partner institutions in a phased manner.

Proposed model for m-learning library services at IGNOU

The model proposes to 'mobilize' existing services so that they will work better with mobile devices. Libraries need to study the opportunities for innovation in the areas of content, systems and tools, services and environments – both physical and virtual – before designing and proposing models for mobile device-based services. Today's distance learners are more versatile in their use of mobile devices and the internet than are older generations. As the university library makes plans to address the needs of its clientele via mobile devices, it is important to determine user requirements and the levels of service required.

First of all, the university library has to identify its clientele. In IGNOU's proposed model, mobile users can be:

- faculty conducting research or teaching activities at university headquarters and regional centres
- research and teaching assistants at IGNOU
- academic councillors at study centres
- student support and other staff at IGNOU
- students studying by distance education at IGNOU
- students whose courses include online and on-campus components
- learners in the research field pursuing doctorates
- learners using mobile devices such as clickers in the classroom
- learners using mobile devices for learning activities outside the study centre classrooms, e.g. class assignments and group projects.

(Lippincott, 2008)

The next step is to identify what m-library users will want to access and read on their mobile devices. A growing number of digital collections and resources are now available in mobile-readable formats. These are typically managed by libraries for flexible learning, and in an institution like IGNOU they will commonly include:

- learning objects supporting courses, usually under copyright of the home institution
- video collections of university courses and archives
- image collections from the university archives
- research paper collections
- research data collections
- theses
- past examination papers
- administrative documents, policies, contracts, etc.
- in-house library databases, e-journals, e-books and data collections
- digitized readings of copyright materials such as journal articles and book chapters supporting courses developed and administered by the university.   (Borchert, 2004)

Besides providing the in-house content, the library will also want to offer licensed content for mobile devices. Libraries may also want to offer a set of mobile formatted reference materials for students studying through an ODL system. In some subject areas, students and researchers rely on quick-reference sources; libraries may wish to consider developing brief guides to reference sources by discipline and then linking to mobile phone-compatible reference sources.

In addition to the above, an initiative may be taken to provide m-library services such as OPAC (Online Public Access Catalogue), SMS alerts, library circulation, RSS feeds, m-blogs, peer-group interaction, reference librarian, access to repositories, and specialized content, including online databases, e-journals and e-books, according to the requirements of library users. When providing links to publishers' websites care is necessary to ensure that they are m-compatible. The content should be accessible by identified groups of users through IDs and passwords.

M-services requirements of IGNOU distance learners: survey results

In order to draw up a model for m-library services at IGNOU, a questionnaire was prepared to assess distance learners' actual needs and expectations for various m-library services. The questionnaire was prepared and distributed personally to learners visiting IGNOU Campus to seek student related services. As it was not possible to distribute the questionnaire to a large number of distance learners scattered around the globe, the survey was limited to a sample in the Delhi region.

Out of 220 users, 132 responded to the questionnaire, a response rate of 60%. The highest numbers of responses were from BCA and MBA students. Table 20.1 shows IGNOU distance learners' responses regarding the various m-library services.

On the basis of survey results and the accessibility of common mobile devices, the m-library services can be offered to IGNOU distance learners in two phases. Phase I will include library OPAC, SMS notification services, mobile instructions, reference/enquiry, RSS alerts, access to m-repository, etc. Phase II will include mobile library databases, links to open sources, moblogging, publishers' databases, mobile library circulation, etc.

Table 20.1 Response rates showing the requirement of m-library services

Type of m-library service required	Response rate (%)
Enrolment status	82
Status of entrance/term-end examinations	82
Assignment evaluation status	85
Alert services about the new programmes/services launched	88
Examination schedule	85
Library alert services about new arrivals	42
SMS reference service	30
Mobile blog posts	25
Access to IGNOU course materials	55
Access to assignments	70
Project reports	60
Past examination papers	90
Webcasting of IGNOU channels	20
Library web-OPAC	60
E-books/e-journals from library	45
Digitized collection of special library materials	55
Reference/enquiry services	60
Library circulation services	30

M-library services in Phase I

Web-OPAC services: Learners can browse the library collection through web-OPAC and can know the availability status of desired documents.

SMS notification services: The library can introduce the SMS service for several functions, such as:

- alerts for enrolment status, examination results
- alerts about new programmes/services
- alerts about new arrivals
- reference services for distance learners
- alert when requested document is available (collect messages)
- reminder when a book/document is due
- loan requests
- document renewals
- requests for account of outstanding fines
- checks for availability of resources
- requests for library opening hours.

Mobile instructions: This includes the use of mobile devices for library instruction, which can be text, audio or video based.

Reference/enquiry: Libraries have been providing reference service by phone for many years, and most university libraries also provide reference services via a range of communication channels, such as chat, instant messaging, texting and e-mail, which is now made easier by mobile phones. The reference desk takes calls from library users in meeting or study spaces in the library. Data from an ongoing study of Virtual Reference Services indicate that even where learners are physically in the library they may prefer to use chat reference rather than seek out a face-to-face encounter. Again, convenience and workflow integration are important.

RSS alerts: RSS is becoming pervasive. Text message and e–mail alerts are also more common. Learners can be told about events, about the status of their interactions/requests and about the availability of resources. RSS feeds, widgets and Facebook applications can also be used.

M-repository: Database of all IGNOU course materials, assignments, project reports and test/examination papers.

M-library services in Phase II

With growing ever-changing mobile technologies and their increasing accessibility to learners, more advanced services can be offered to our distance learners in Phase II. These services are:

Mobile library databases: These include digitized collections of selected books, research papers and institutional documents, abstracts of journals and periodicals, etc. Distance learners would turn on their phones and click on a library icon that offered them shortcuts to desired library content such as the OPAC, e-books and audio books, without ever having to open a web browser. This is already possible now, with the proper programming.

Links to free Open Access resources: The library can compile lists of links to open access materials and make them available on the web. The service can also be provided on demand through the mobile network.

Moblogging: A moblog is a mobile blog that is maintained via a mobile device such as a PDA, mobile phone or mobile PC. The widespread adoption of camera and video-enabled phones has driven the moblogging trend in education. Distance learners can publish blog entries directly to the web from a mobile phone or other mobile device. For faster communication, Twitter can be used to help learners to post write-ups directly from their mobile phones, even when on the move, and to interact with library staff and their peer groups.

Publishers' databases: The university library may subscribe to databases that the different groups of users can use, or it may house them in the institutional repository. These include e-books, audio books, e-journals, etc., which can be used on mobile devices. The library will subscribe to the specific m-compatible resources. The collections can be downloaded from the library website to users' own mobile devices. Libraries can provide authentication mechanisms for downloading to mobile devices.

Mobile presentation and visibility: Videos and podcasts describing or promoting particular library services, covering library events and so on are becoming more common. Often, these are made available on network-level sites, such as YouTube and iTunes, where they are more visible.

Mobile library circulation: A hand-held circulation tool like PocketCirc can be used, which enables library staff to access the library management system on a PDA device. This wireless application enables staff to help learners in the stacks, check out materials while off site, e.g. at community or campus events, and update inventory items while walking around the library.

Video conferencing: Distance learners can click on a mobile phone icon which will initiate a video conference with the relevant library staff for a number of reference services. With powerful services such as Skype Mobile, this has now become a reality. (Kroski, 2008, 46)

Web layout of m-library website

Today, library websites are being transformed into m-library websites. Either a new m-website can be created or an existing site can be transcoded. Transcoding is a technology which takes a regular website and reformats it for display on a mobile screen. When using a mobile device, many search engines, including Google, will show transcoded versions of web pages as results, along with any mobile editions of the site. Many free transcoding applications, such as Skweezer and Mowser, are available, which compress the HTML content of a website to produce a single-column, spartan version of the original that can be viewed via a mobile browser.

Transcoded web pages are viewable on a wide range of mobile phones. However, the automatic nature of the web page transformation often results in excessive scrolling and less-than-perfect displays. Hence, designing a mobile-specific website not only provides more freedom in the design, content and structure of a portable web page, but also allows for additional decisions about the type of technology and format in which to develop it. It is therefore advisable to create more effective m-web pages. The following suggestions might be considered while designing a library mobile website:

Keep it simple: 97% of users don't have QWERTY keyboards on their phones; as of the end of 2010 only 38% of phones will be 3G enabled. Keeping this limitation in mind, m-library websites should be simple and easy to use.

Test on various platforms: Mobile websites may display differently on various browsers and phones, therefore it is advantageous to test m-library web pages on as many different devices and operating systems as possible. Various freely available web-based emulators can be used for testing.

Keep customization on the desktop: Enable learners to perform various activities such as setting up their user profiles, favourites/bookmarks and adjusting preferences from their personal computers, as this is too time consuming and difficult on a mobile device.

Incorporate search: Due to the difficulty of navigating the mobile web, adding a search option near the top of an m-library website will save precious time for distance learners and avoid excessive scrolling on their handsets.

Remember usability: Information architecture is more vital on the mobile web. The key contents should be placed at the top of the page so that distance learners do not have to navigate further to find crucial items.

Clean up images: Due to the small screens of handsets and the expense of downloading to mobiles, use of unnecessary and decorative images should be minimized on mobile websites.

More mobile-suitable content: Attempts should be made to make use of existing content and services for mobile use, as well as looking at ways of developing new m-library services for distance learners on-the-go.

A new programming language called XHTML-MP allows developers to create robust mobile websites, but it is only viewable on newer, high-end devices, while the older, less feature-rich WML (Wireless Markup Language) is compatible with mass-market phones, making it the safer choice.

Many free applications are available to help organizations to create their own mobile websites, such as Winksite and Mofuse, which create mobile versions of websites from RSS feeds. These programs also (a) provide tools to create Quick Response barcodes, and widgets that can be added to desktop websites and used to send the mobile URL to visitors who enter their phone numbers, and (b) embed code for adding the site to blogs or other websites and iPhone-only websites, and links to share websites with social networks and communities. Applications such as Zinadoo and dotMobi's Site Builder provide FrontPage-like development interfaces for creating mobile websites from scratch (Kroski, 2008, 52).

IGNOU Library needs to prepare an authenticated mobile interface for its website. The library website needs to be designed so that it can be viewed on mobile devices and needs to be much simpler than the typical library site. It will enable users to browse the library website while on the move. Several libraries around the world are now providing mobile library interfaces and applications. The following are some examples that can be referred to when designing a mobile-optimized website for IGNOU Library.

Athabasca Open University has implemented a comprehensive m-library website. In this m-library system users' devices can be auto-detected and an appropriately formatted mobile version of the website is presented. The library website has been implemented in PHP. In real time, the different browsers are detected by a server-side script. By parsing the HTTP_USER_AGENT, the server can identify the platform, whether it is Windows CE or Palm OS. The system chooses the correct style-sheet and display model. Digital information is reformatted on the fly for different browsers and screen resolutions. The m-library website provides a wide range of digital resources and library services, including the digital reading room (DRR) or e-course reserve, the digital reference centre (DRC), the digital thesis and project room (DTPR), a help centre, a search engine, journal databases, AirPAC (a mobile library catalogue application) and library services through the world wide web (Yang et al., 2006).

Open University of Catalonia carried out a pilot test with SMS in 1999. The university managed to link up mobile phones to the OPAC through a WAP system that allowed catalogue searching by categories and finding the location of a document, and that also provided the address of a library from which the user could borrow it. The university is also working towards adapting services to a wide range of m-devices used by library users. The library is also experimenting with some new devices, such as e-books, and some new services, to get more feedback through the OPAC and meta-library search products (Pérez, 2009).

Some other m-library website models are:

■ American University Libraries[1]
■ Boston University Medical Center Mobile Library[2]
■ University of Illinois Library[3]
■ New York University Libraries[4]
■ University of Virginia Library.[5]

In view of the above, the screen shot from the IGNOU Library website (Figure 20.1) has been reformatted for optimum mobile performance and automatically added with additional mobile specific features such as 'click-to-call' and Google Maps using Mowser online software, which allows you to view the web through your mobile phone.

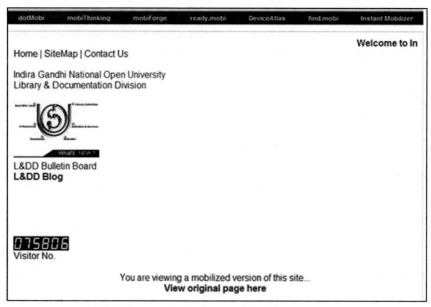

Figure 20.1 IGNOU library website formatted for mobile access

Conclusion

M-libraries offer a unique opportunity for distance learners in a range of settings. They enhance flexibility for distance learners; however, they demand new pedagogies and new approaches to the delivery of the library services. In a country such as India, where most of the population reside in remote and rural areas, there are a number of challenges for the effective implementation of m-library services. The availability of 3G mobiles, and awareness and cost-consciousness on the part of distance learners are some of the major challenges for information professionals. Apart from this, connectivity of mobile signals and other limitations of mobile devices related to hardware and software characteristics and the limited display capability of the tiny screens of mobiles have to be considered when designing web content for small-screen devices. Hence, if appropriately facilitated, m-libraries can help learners greatly by providing virtual library services on their mobile devices. Library professionals need to spend time and effort preparing and delivering a variety of rich learning contents and resources, along with other interactive library services. This in turn, will contribute to the quality of the distance learning process. To keep up with the changing environment and to continue to effectively facilitate m-library services, librarians and information professionals need to learn about new methodologies and adapt to the changing environment when and where appropriate. Mobile web services provide a means of integrating applications via the internet. The proposed m-library services architecture could provide a new direction for developing the m-library website at IGNOU Library, hence strengthening m-learning in distance education systems.

Notes

1   www.library.american.edu/mobile/
2   http://med-libwww.bu.edu/mobile/index.cfm
3   http://hades.grainger.uiuc.edu/nikki/Mobile/Version1/index.htm
4   http://library.nyu.edu:8000/mobile/
5   http://mobile.virginia.edu/library.php

References

Arora, S. K. (2007) *Concept note NODLINET*,
    www.ignou.ac.in/divisions/library/N-About.htm
Borchert, M. (2004) Integrating Library Content into the University's Online
    Learning Environment, *Flexible Learning and Access Services*,
    www98.griffith.edu.au/dspace/bitstream/10072/2427/1/Borchert_M_eAgenda_
    2004.pdf [accessed 20 August 2009].
IGNOU (2008) All Times Everywhere, *IGNOU Foundation day Special*, 2008, 35.

IGNOU (2009a) Indira Gandhi National Open University, *IGNOU Brief*,
www.ignou.ac.in/aboutus/ignoubrief.htm.

IGNOU (2009b) Indira Gandhi National Open University Profile 2009,
www.ignou.ac.in/profile_09/profile-2009%20(Final).pdf.

Kanjilal, U. (2008) Reaching the Unreached with e-Learning, *Digital Learning*,
(May)
www.digitallearning.in/articles/article-details.asp?articleid=1933&typ
=COVER%20STORY [accessed 20 August 2009].

Kroski, E. (2008) On the Move with the Mobile Web: libraries and mobile
technologies, *Library Technology Reports*, 44 (5), 6–7, 46 and 52.
http://eprints.rclis.org/15024/1/mobile_web_ltr.pdf [accessed 20 August 2009].

Lippincott, J. K. (2008) Mobile Technologies, Mobile Users: implications for
academic libraries, *Association of Research Libraries, Bimonthly Report*, No. 261,
(December),
www.arl.org/bm~doc/arl-br-.261-mobile pdf [accessed 20 August 2009].

Pérez, D. (2009) M-Library in an M-university: changing models in the Open
University of Catalonia,
http://m-libraries2009.ubc.ca/Abstracts/19Abstract.pdf [accessed 20 August
2009].

Siliconindia News Bureau (2008) India's Mobile Base to Reach 500 Million by
2010, (5 December),
www.siliconindia.com/shownews/Indias_mobile_base_to_reach_500_Million_by
_2010-nid-49588.html [accessed 20 August 2009].

Suresh, P. V. (2008) Education Gets Mobility in Digital Learning, *Digital Learning*,
(May)
www.digitallearning.in/articles/article-details.asp?articleid=1925&typ=
M-LEARNING [accessed 20 August 2009].

Yang, C. et al. (2006) The Athabasca University Mobile Library Project: increasing
the boundaries of anytime and anywhere learning for students, *International
Conference on Communications and Mobile Computing, 2006*,
http://delivery.acm.org/10.1145/1150000/1143808/p1289cao.pdf?key1=
1143808&key2=6333611421&coll=GUIDE&dl=GUIDE&CFID=
32792016&CFTOKEN=65884350 [accessed 20 August 2009].

21

Bridging the mobile divide: using mobile devices to engage the X and Y generations

Phil Cheeseman and Faye Jackson

Introduction

At Roehampton University, London we have implemented a range of flexible learning spaces, both formal and social. The university library has been a focus for these developments and we aim to take full advantage of the potential offered by new and emerging technologies.

We have undertaken a study to look at the impact of ubiquitous computing on approaches to study and expectations of library environments. In this case study we describe how the use of mobile devices has enabled us to engage schoolchildren as well as our own students as participants in the future planning of our library and its services. We will describe how this initiative enabled us to extend the reach of the library and its services and how it subsequently led to the development of a social learning space for staff to explore the use of mobile technologies.

Schools surveys

The study was undertaken as part of an outreach event with schoolchildren aged between 10 and 15 years from several schools local to the university. Using audience response systems (clickers), audio recorders and flip video recorders we asked the groups first to take part in a survey to describe their study habits, how and where they liked to study and to answer questions about resources and what technologies they used in their social time.

Results showed that many preferred to sit on their bed and listen to music while doing their homework or to lounge in front of the TV, but equally some wanted to sit at a desk to study. This is in line with results from our university library student survey, and reaffirms the need for us to provide a variety of study environments. When asked where they would most frequently go for information to help with homework, the children responded resoundingly in favour of Google, followed closely by Wikipedia. In relation to communicating with their friends, they preferred instant messaging or phone for instant access to each other. Some were members of social networking sites and used them as communication tools, but these were generally far fewer.

We then asked the schoolchildren to draw, plan and list the features that they would like to have in their ideal library/learning centre and present them to the group. They were very interested in design features and their surroundings and often incorporated their current study and social environments into their drawings. Refreshments, comfortable furniture and social areas featured prominently and frequently included TV/DVD and music areas in chill-out zones with games consoles and recreational activities. They also included books and lots of computers and quiet space for studying. Overall, they wanted a comfortable environment that was colourful and inspiring, with ready access to technologies and people on hand to help. They were also very creative in their designs, including spaces for displaying artwork, areas for presenting events and performances, beds for taking a break, and 'brain-booster bars' (see Figure 21.1). In particular, some of the technology-based creations were fun and inventive, including the 'MyPod'.

> We've got the MyPod, it's like your bed, you fall asleep in it and in your sleep learning flows into your brain and its only £15 milllion.

The Green Room

The enthusiasm for technology demonstrated by the schoolchildren and by our own students suggested a disparity between student expectations and staff knowledge and skills. Prompted by this, we have developed a new social learning space for staff called the Green Room. This resource, which opened in September 2008, provides an environment for staff to explore and experiment with learning technologies, including a range of mobile devices, to develop strategies for using them and to share good practice with one another. Through exchange of ideas, reflection on

Figure 21.1 Creative library design

individual practice and collaborative development, the activities undertaken in the Green Room provide a 'sense of place' for the existing e-learning community, as well as a point of introduction for new members of staff.

In recognition of the value of active and social approaches to learning, the room provides a test ground for exploring the benefits and challenges of using learning technologies. A number of mobile devices, including smart phones, Flip video cameras and audience response systems (clickers) can be borrowed by staff for use with students. Initial experimentation leads to pilot projects, and from there to embedded practice.

In one such project, students studying to be teachers used GPS-enabled smart phones in geography fieldwork. During the exercise pictures and video taken with the phones were GPS-tagged and attached to Google Maps. In another, research students accompanied staff on a trip to a refugee camp in Western Sahara to run workshop activities as part of a film festival. They used Flip video cameras and digital cameras to provide a means for the Saharawi people to tell their story, opening up a channel of communication with the international community.

Conclusions

Use of a range of mobile devices has enabled us to engage with schoolchildren and university staff in two projects that, though distinct from one another, have obvious synergies. Exploring the study habits and expectations of our future students has highlighted not only the importance that they attribute to a variety of learning spaces but also the ease with which they accommodate new technologies and assume that they will be available to them.

One year on from opening the Green Room, we have seen that providing a comfortable, safe space for staff to explore technologies can be an effective strategy for addressing the disparity between student expectations and staff knowledge and skills. Comments left by staff in our visitors' book speak for themselves.

Returned voting system – students loved them!! Thanks.

Great space – valuable tools for brainstorming and group meeting.

Creative writing woz here – we think it's great! Buzzing with ideas and applications.

22

Information literacy gets mobile

Peter Godwin

The pervasiveness of mobiles gives a great new opportunity for librarians to reach their users. Within the context of the changing landscape mapped by Google generation reports (CIBER, 2008), information literacy (IL) remains a crucial deficit area. The use of mobiles for IL could be seen as an extension of Web 2.0, e.g. podcasts and screen casts. These are becoming mainstream for the new librarian. Adoption of mobile devices in libraries is in its early days because of the limitations of the technology and the variety of devices owned. There are serious limitations, and it is time to experiment. This chapter discusses the potential, looks at present practice and suggests ways forward.

The mobile background

By the end of 2008, there were over four billion mobile subscriptions worldwide, representing 61% of the population, and there were 335 million broadband subscribers. By contrast, just under a quarter of the 6.17 billion world population use the internet (International Telecommunication Union, 2009). Mobile devices are so widespread that the most common items we carry around with us are keys, money and the mobile phone. The ECAR study 2008 in the USA (Salaway et al., 2008) found that 66.1% had internet-capable mobile phones, and of these 25.9% accessed the internet once a week or more. In 2007, 14% accessed the internet on a mobile device on a typical day and 32% had never used the internet on a mobile at any time. This had changed in 2009 to 23% and 38%

respectively (Pew Internet & American Life Project, 2009). As McIntosh (2008) said 'it's about time schools sharpened their focus on how they can help students power up their learning with their mobile devices, rather than have them power down at the school door.' I believe the same is true for libraries.

Another way of engaging?

Much has been written about finding ways of engaging with the web generation. The CIBER report has transformed this debate and brought the realization that we are all part of this generational change and exhibit some of its so-called major characteristics, such as preferring to learn via small chunks of information (CIBER, 2008). Crucially, it found that not all users are equally technically proficient – geeks are in the minority, and the lack of IL is a real concern. Gross and Latham (2007) suggested that those with low-level IL skills often had inflated views of what they knew, and that innovative ways would be needed in higher education to address this. The CIBER report showed that everyone likes to learn by small chunks, rather than by reading long and complicated manuals.

Just as Web 2.0 has given librarians the possibility of using tools that encourage active learning in the place where the student is, so can the use of mobiles. A ground-breaking report by Toni Twiss in New Zealand (Twiss, 2008) that examines the use of mobiles for IL in schools found 'the potential for mobile phone use in the classrooms exciting'. One primary and two secondary classes were surveyed using Vodaphone 3G phones with OperaMini to view the web. It was found that the best use was for small, guided tasks within the lessons. Many students did struggle with the navigation, because not all were equally tech savvy. There were problems with speed, screen size and ease of use. Interestingly, the primary pupils were the most enthusiastic. At the other end of the educational spectrum, Char Booth, in her survey of take-up of Web 2.0 at Ohio State University (Booth, 2009), found that the postgraduates were the most likely to use tools and material via mobiles. Clearly, we have not yet moved beyond the experimentation stage.

What are we trying to deliver?

'Information literacy is knowing when and why you need information, where to find it, and how to evaluate, use and communicate it in an ethical manner' (CILIP[1]). This is not the place to have a discussion on which framework to use to deliver this, and I am taking the SCONUL 'Seven Pillars'[2] as my starting point . In practice, along with all such frameworks, this is likely to produce the MEGO (my eyes glaze over) effect. Librarians have the possibility of using mobiles to counter this, along with Web 2.0 tools to reach at least some of their clientele. This will require them to change their attitude and behaviour toward mobiles in libraries, harnessing them rather than banning their use.

Which devices should be used?

It is important to recognize that the variety of mobile devices presently available poses problems. PDAs, mobile phones, Blackberries, iPods, iTouch iPhones and other high-level phones exhibit different characteristics and give different possibilities. Therefore, special mobile pages may be deemed desirable; there may be a lack of additional windows on a mobile, poor navigation, no JavaScript or cookies available; some websites may be inaccessible (e.g. videos, pdfs), with slow access speeds, pages broken up or compressed; and finally there can be the high cost of access. Despite these issues, I prefer to see this as a transitional stage and a time for experiment. The iPhone continues to collect a lot of hype, and sales have held up very well in the recession. It has heralded a generation of advanced web-capable mobile devices. This has already begun with the Google Android and the PalmPre. The challenge for librarians is to decide whether to implement fixed-term projects for higher end phones, which most students don't yet have, or whether to concentrate on services for lower end mobiles. I guess the answer is both!

How could IL become mobile?

The following discussion covers both delivery of information via mobile devices and new tools that could be used in IL delivery. Some of these (e.g. mobile interfaces for websites) could also prove to have promotional benefits. Users who have difficulty knowing where to go for information, or how to go about it, are now offered access any time, any place by using their mobiles. 'This is the yet unrealized potential of computing ubiquity and the library' (Hahn, 2008).

Mobile sites

Providing a mobile interface of an existing library website was a starting point for some libraries, e.g. Mobilib at North Carolina State University Library. This offered the potential of new outreach and a way of engaging users. Some other notable libraries that have taken this route are Duke University, and Washington DC Public Library. Databases are sometimes now being given special mobile-friendly versions that are easier to navigate and access on mobiles e.g. Westlaw and PubMed. These could ease our IL mission by giving users another route to our databases. Use of a Twitter special account for communicating with a special group, or simply as a library site for current awareness, tips, information flow, reference help and instruction and teaching IL online is being tried at Santa Barbara City College, California (Neufeld, 2009). RefWorks is now available on mobiles using RefMobile on a smart phone, mobile or PDA. Loyola University in Chicago, for example, is already recommending this to its students. Referencing is a key way of demonstrating the sources used, and therefore of promoting ethical use of sources as part of IL.

Tours

Tours are beginning to use mobile devices rather than proprietary devices (often used in museums). Audio tours of the Headington and Wheatley sites at Oxford Brookes University[3] can be downloaded for use on any MP3 player. Temple University in Philadelphia is offering mobile phone tours.[4] As Gretchen Sneff, head librarian for Science, Engineering and Architecture said, 'Students always seem to have their cell phones with them. Offering the self-guided audio tour via cell phone enables us to provide library users with information they need when they need it – even when there is not staff on duty.' Students call a number and enter a tour stop number followed by the # key, with no charge for the call, just usage of the mobile minutes, and there is provision for feedback.

Reference

Use of SMS texting by reference desks is being tested in many libraries, and could be regarded as IL individual help or tuition. Library Success: a best practices wiki[5] contains a long list of libraries (mainly in the USA) offering SMS reference services. Joe Murphy, Science Librarian, Coordinator of Instruction and Technology, Yale University Science Libraries co-ordinator, is a trendsetter and great proponent of text referencing services in academic libraries. He believes the future of libraries lies in the total integration of services and collections with mobile devices via SMS, Twitter and other mobile applications (Murphy, 2009). A project with librarians using hand-helds for reference service support has been piloted at Penn State University. Librarians used various hand-helds to provide support on the campus and, although all had their advantages, the Fujitsu Lifebook was the preferred device. Small screen sizes did make it hard to view all options available and each device involved a considerable learning curve. The main advantage was the extra dimension it gave to the on-campus roving help with a variety of online tasks (Cahoy et al., 2008).

Screencasting

Screencasting, podcasting and vodcasting are becoming more common in the delivery of IL by librarians, and their use on mobile devices is a natural extension. As already noted, users like to learn in short bursts, whenever and wherever they choose, so the creation of these short broadcasts gives us a great opportunity. The challenge will be in their promotion and making them available in multiple places on our websites, so that users keep finding them. Examples include Hannon Library at Southern Oregon University at Ashland, Oregon and a series at Washington State University. Arizona State University (ASU) has a Faculty Workshop series on the ASU Library channel on iTunes.

Polling

There is increasing interest in the use of Personal Response Systems (PRS) to test comprehension of lectures or gather opinion in classrooms. Poll software at various skill levels is available for use on many hand-held devices, which could be used instead of Keepad clickers. Poll Everywhere is a notable example and was used by Toni Twiss during her project in New Zealand (Twiss, 2008). This enables class participants to choose simple options which are then collocated on a web browser for display on a presentation screen or for later use. Schools in the USA that cannot afford expensive clickers are experimenting with Poll Everywhere as a cheaper alternative, using student mobiles (Learn, 2009).

QR codes

The use of QR codes is widespread in Japan. They enable text or a phone number or a URL to be encoded and given a special two-dimensional barcode. There are a number of free services for doing this on the web, e.g. BeeTag and Neoreader. Most phones in Japan have a QR reader already installed, but otherwise users need to download free software, e.g. BeeTag, Neoreader, to their phones. They can then scan any QR code and go straight to the object. The advantage of using these is that users don't have to copy a long URL, and can instead simply capture it and go to the object. Text and phone numbers are the easiest to access for all mobile users; URLS will require Wi-Fi access. At present, knowledge of QR codes is weak in higher education, as can be seen from investigations at the University of Bath (Ramsden, 2008); Andrew Walsh describes the possibilities and present limitations for libraries (Walsh, 2009). Theoretically a QR code displayed on a shelf or wall next to the printed run of a journal could link into the online full text. In practice, we should expect this to take a while to display. Another suggestion might be to include a QR code in a teaching handout or display it in the library where it would most help the user (e.g. beside a self-issue machine). The user could then point their mobile at the code and play the relevant video. We shall be testing this in Learning Resources at the University of Bedfordshire. YouTube can be watched on the higher-end mobile devices and this format is ideal for short tutorials. As with screencasting, it will rely on heavy promotion to have any impact.

Conclusion: where next?

The variations and differing capabilities of mobile devices at present make generalization difficult. Speed is problematic and the cost of devices and of access even more so. Tests in the recent Nielsen report on mobile usability concluded that 'designing for mobile is hard. Technical accessibility is very far from providing an acceptable user experience. It's not enough that your site will display on a phone. Even touch phones that offer "full-featured" browsers don't offer PC-level usability in terms of users' ability to actually get things done on a website' (Nielsen, 2009). I believe it reasonable to predict that access prices will come down. 'Smartphone sales were strong during the second quarter of 2009, with sales of 40.9 million units in line with Gartner's forecast of 27 per cent year-on-year sales growth for 2009' (Gartner, 2009). We are seeing that the mobile revolution is not one of the areas being squeezed hardest by the world recession. We will need to see more Bluetooth and joined-up Wi-Fi access to really spur the kind of developments I have been suggesting. I believe mobile devices will be significant for libraries because 'the nature of search in this new world of mobile internet devices will shift. This is because the journey that Generation Y is taking on the Internet is more concerned with social expression than finding information' (Taptu, 2008). Be warned that libraries have to look to the social aspects of communication in their promotion and delivery. Mobiles can foster active learning techniques for varied learning styles. They are not a panacea; rather, another way of communicating our messages and services. Orange County Library System, Orlando, Florida[6] is showing us the way. We should not underestimate the speed of change in this area. The IL connection is still young, so this is the time to pioneer.

Notes

1   CILIP Definition of Information Literacy,
    www.cilip.org.uk/policyadvocacy/learning/informationliteracy/definition/.
2   SCONUL Seven Pillars of Information Literacy framework,
    www.sconul.ac.uk/groups/information_literacy/seven_pillars.
3   Oxford Brookes University Audio tours at Headington and Wheatley,
    www.brookes.ac.uk/library/tours-audio.html.
4   Temple University Library,
    www.temple.edu/newsroom/2008_2009/04/stories/cellphones_library.htm.
5   Library Success: a best practices wiki,
    www.libsuccess.org/index.php?title=Online_Reference.
6   Orange County Library,
    www.ocls.info/mobile/video.asp.

References

Booth, C. (2009) *Informing Innovation: tracking student interest in emerging library technologies at Ohio State University*, ACRL,
    www.ala.org/ala/mgrps/divs/acrl/publications/digital/ii-booth.pdf.

Cahoy, E. S. et al. (2008) *Handheld Devices Pilot Project*,
www.libraries.psu.edu/etc/medialib/psulpublicmedialibrary/lls/
documents.Par.18154.File.dat/HHDFinalReport.pdf.

CIBER (2008) *The Information Behaviour of the Researcher of the Future*, Report
prepared for the British Library and JISC,
www.bl.uk/news/pdf/googlegen.pdf.

Gartner (2009) *Gartner says Worldwide Mobile Phone Sales Declined 6 per cent and
Smartphones Grew 27 per cent in Second Quarter of 2009* (Press release),
www.gartner.com/it/page.jsp?id=1126812.

Gross, M. and Latham, D. (2007) Attaining Information Literacy: an investigation
of the relationship between skill level, self-estimates of skill, and library anxiety,
Library & Information Science Research, **29** (3), 332–53.

Hahn, J. (2008) Mobile Learning for the Twenty-First Century Librarian, *Reference
Services Review*, **36** (3), 272–88.

International Telecommunication Union (2009) *Measuring the Information Society*,
www.itu.int/ITU-D/ict/publications/idi/2009/material/IDI2009_w5.pdf.

Learn, N. C. (2009) *Use Cell Phones to Poll your Students*. Instructify (blog),
http://instructify.com/2008/07/18/use-cell-phones-to-poll-your-students.

McIntosh, E. (2008) *Please Turn your Mobiles On*. BBC Scotland Learning (blog),
www.bbc.co.uk/scotland/learning/diary/mobile.shtml.

Murphy, J. (2009) Sending out an SMS. *Handheld Librarian Online Conference, 2009*,
http://vodpod.com/watch/2008380-handheld-librarian-online-conference-
sending-out-an-sms.

Neufeld, K. (2009) Moodle, Facebook, Twitter and Teaching Online. Misc.joy
(blog),
http://kenleyneufeld.com/2008/04/18/moodle-facebook-twitter-and-teaching-
online/.

Nielsen, J. (2009) *Mobile User Experience is Miserable*. Jakob Nielsen's Alertbox,
www.useit.com/alertbox/mobile-usability.html.

Pew Internet & American Life Project (2009) *Internet Wireless Use – July 2009*,
www.pewinternet.org/~/media//Files/Reports/2009/Wireless-Internet-Use.pdf.

Ramsden, A. (2008) *The Use of QR Codes in Education: a getting started guide for
academics*,
http://opus.bath.ac.uk/11408/1/getting_started_with_QR_Codes.pdf.

Salaway, G. et al. (2008) *ECAR Study of Undergraduates and Information Technology
2008*,
www.educause.edu/ers0808.

Taptu (2008) *Unleashing Search for the Mobile Generation*.Taptu white paper,
www.scribd.com/doc/12748128/Unleashing-search-for-the-mobile-
generation2008.

Twiss, T. (2008) *Ubiquitous Information: an eFellow report on the use of mobile phones in
classrooms to foster information literacy skills*,
www.scribd.com/doc/9507014/Toni-Twiss-Ubiquitous-Information.

Walsh, A. (2009) Quick Response Codes and Libraries, *Library Hi Tech News*, 5/6,
7–8.

23

Library and Student Support (L&SS): flexible, blended and technology-enhanced learning

Victoria Owen

Introduction

In recent years there has been growing debate about a new method of teaching and learning which is flexible, blended and facilitated by technology, allowing 21st-century learners to work and study in more individual, personalized and portable ways. This new era of learning is known as mobile learning (m-learning).

The m-learning concept has been pushed to the forefront of educational debate by the present ubiquitous ownership of mobile devices amongst learners (such as mobile phones, portable audio players and laptop computers), which has created a perceived need amongst this generation of learners to use such devices in the academic arena. Furthermore, employers are indicating that the skills sets that graduates currently possess are no longer relevant in the present working world; the ability to collaborate, communicate, be creative and constructively criticize are now high on the agenda (Bruns et al., 2007).

Background

Within Liverpool John Moores University (LJMU) there are key factors influencing the requirements for more flexible, blended and technology-enhanced learning opportunities. In 2006 the LJMU Academic Board approved a revised and updated Learning, Teaching and Assessment (LTA) Strategy. New and/or enhanced priority areas include a greater focus on student skill development as demanded by employers, accessibility and inclusivity, and a research-informed curriculum. M-learning falls under two of these new and/or enhanced priorities, namely key objectives 3 and 4. Key objective 3 stipulates that, as a university, LJMU will support the learning needs of all students where a flexible approach to learning is defined as a key goal. Furthermore, key objective 4 states that LJMU will strive to develop and enhance its student's employability, achieved through connecting students to the world of work, equipping them with the standard base of graduate skills and teaching them digital literacy skills.

A major cross-university review of how it supports the student experience has stimulated a radical review of services, with LJMU's three Learning Resource Centre (LRC) sites adapted to offer improved services to the student body. Technology-rich and flexible learning spaces will be two key enhancements to the learning spaces currently provided, in which student use of personal and mobile devices will be supported and encouraged. To address this shift, Library and Student Support (L&SS) carried out a six-month research project to critically investigate and analyse m-learning in a higher education (HE) context and examine how it can best be supported.

Methodology

The research followed a case-study model allowing for the development of detailed and intensive knowledge of student/staff m-learning attitudes within LJMU and the collection of data via a range of data-collection techniques.

Two online questionnaires were distributed, one accessible by the entire LJMU body via the L&SS web pages and one distributed specifically to those students studying on distance learning courses. Face-to-face student interviews were performed at the three LRC sites, mainly with on-campus learners. Focus groups were used to collect more in-depth qualitative data; the student focus group expanded on the questions of the face-to-face student interviews; the staff focus group explored support issues entailed within an m-learning culture.

Results

The mobile phone was the most commonly owned device amongst the LJMU student body, while there were also high ownership levels for laptops and audio listening devices. Netbooks, hand-held gaming devices and smart phones were owned by a few, whilst PDA ownership was negligible. Students' use of mobile devices was wholly traditional; for example, the mobile phone was mainly used socially, for talking, texting, playing music and taking photos; audio listening devices were used recreationally, to listen to music, with a small number listening to podcasts.

Mobile device use for teaching and learning activities was only notable with laptops; however, the majority of laptop usage was in traditional academic activities such as researching and writing assignments in a setting that is indistinguishable from that for a desktop PC or MAC. It appeared, however, that LJMU students *would like* to be able to access common technologies (such as Blackboard, LJMU Electronic Library, e-mail) in more mobile contexts and that there was a distinct need for teaching and learning materials to be provided in more mobile ways. There was some indication of a lack of coherence with regard to the m-learning objects provided as part of current learning experiences at LJMU.

The major theme that emerged from the staff perspective was difficulty in providing support for the vast number of mobile devices, operating systems and platforms owned by the LJMU student body. However, it was deemed necessary that a baseline level of support should be agreed and advertised in order that, as an institution, continuing high levels of support and technical expertise could be fulfilled. This baseline could be further developed and expanded in parallel with the growth in knowledge and expertise amongst support staff members within LJMU.

Conclusions

It is apparent that 21st-century learners studying at LJMU require more flexible, blended and technology-enhanced learning opportunities. LJMU students value the opportunities offered by technology in relation to their learning experience and appear to be dissatisfied with the current access routes available for common technologies (such as Blackboard, LJMU Electronic Library and e-mail). There is a disparity between the learning materials that LJMU students would like to exploit in more mobile contexts and those that they currently can. The absence of a coherent approach to m-learning was evident. The lack of coherence in provision of m-learning materials carries over to support of m-learning. At present there is no clarity as to which devices are supported and which are not, making it difficult for support staff to develop their skills in this area in a relevant manner. However, support staff continued to assist students with their mobile devices as far as was possible.

References

Bruns, A., Cobcroft, R. C., Smith, J. E., Towers, S. J. (2007) Mobile Learning Technologies and the Move towards 'User-Led Education', *Mobile Media*, (2–4 July), http://eprints.qut.edu.au/6625/.

Part 5

Building the evidence base for m-libraries

24

Enhancing library access through the use of mobile technology: case study of information services provided by six mobile companies in Bangladesh

Nafiz Zaman Shuva

Abstract

While mobile technology is readily available in Bangladesh, as in many other developing countries, innovations in its use for library and information centre service development have surprisingly not started yet. This chapter explores the prospects of using mobile technologies for information services in Bangladesh. The research explores the existing information services of six mobile phone companies of Bangladesh. The chapter shows the current status of information technology at the various library and information centres of Bangladesh. The research identifies some library and information services that could be offered using mobile technology. Recommendations are made as to how mobile technologies might be used effectively in designing various information services and satisfying the information needs of various users.

Introduction

Bangladesh emerged as an independent, sovereign country in 1971 following a nine-months war of liberation. It has one of the largest river deltas in the world, with a total area of 147,570 sq km. Bangladesh has a population of about 140 million, making it one of the most densely populated countries of the world. The literacy rate amongst the population is 43.1%. Over 98% of people speak in Bengali, but English is widely spoken (National Web Portal of Bangladesh[1]).

Research on library development in Bangladesh shows that before mid 1800s most libraries were privately owned and their use was limited to certain groups. The library movement started with the establishment of four public libraries in the district towns of Jessore, Bogura, Barisal and Rangpur in 1854 (Shuva, 2005). Subsequently, libraries were established in other district towns. After the war of liberation in 1971 library development increased substantially, with many public and special libraries being established after independence.

Status of ICT in different types of libraries in Bangladesh

Computers were first introduced into Bangladesh in 1964, with the installation of an IBM 1620 computer at the Atomic Energy Commission (BASIS, 2005). Subsequently, use of computers was introduced in the Institute of Statistical Research and Training (ISRT) in 1964; Bangladesh University of Engineering and Technology (BUET) in 1968; Janata Bank in 1969; Adamjee Jute Mills Ltd in 1970; and the Bureau of Statistics in 1973.

The 1980s are considered to be the beginning of the automation era so far as libraries and information centres in Bangladesh are concerned. The International Centre for Diarrhoeal Disease Research, Bangladesh (ICDDR-B) Library and Agricultural Information Centre (AIC) are pioneers in creating bibliographic databases in specialized fields, using microcomputers (Khan, 1989). Now a good number of libraries are automating library services and offering ICT-based services to their users.

The internet came to Bangladesh with UUCP e-mail in 1993, and IP connectivity in 1996. In mid-June 1996 the VSAT base data circuit was commissioned for the first time. In 2006 Bangladesh connected to the SEA-ME-WE 4 Submarine cable. Following that, many ISPs connected to the submarine cable via the Bangladesh Telegraph and Telephone Board (now BTCL).

Use of computers and the internet in Bangladesh has now dramatically increased. However, with very few exceptions, most libraries are unable to harness the benefits of ICT for the organization and management of information. Table 24.1 shows the existing status of ICT and the possibility of m-library development in different types of libraries in Bangladesh.

It is clear from the table that the ICT status in both government and non-government public libraries and in the National Library of Bangladesh is quite unsatisfactory. The ICT status of college libraries is also very

Table 24.1 Existing status of ICT and possibility of m-libraries development in libraries in Bangladesh

Type of library	Number		Availability of online databases	Availability of integrated library automation system	Internet service available for users	Possibility of m-library development
Academic libraries	University library	31 (public)[a] 53 (private)[b]	Only 15 public university libraries and 14 private university libraries subscribed to online databases through Bangladesh INASP-PERI Consortium	Most of the leading public and private university libraries use integrated library automation system	Yes	Yes
	College library	1750[c]	No	No	Limited	No
Public libraries	Government public library	68[d]	No	No	No	No
	Non-government public library	1603[e]	No	No	No	No
National Library of Bangladesh	National Library of Bangladesh	1	No	No; planning to introduce soon	No	No
	National Health Library and Documentation Centre	1	Online database searching and retrieval service available	No	Yes	Yes
Special libraries	Around 1000*	Only 9 special libraries subscribed to online databases through Bangladesh INASP-PERI Consortium	A good number of special libraries use some kind of integrated library automation system	Yes	Yes	

Sources: [a]University Grants Commission of Bangladesh (2009a); [b]University Grants Commission of Bangladesh (2009b); [c]National University of Bangladesh (2009); [d]Department of Public Libraries (2009); [e]Bangladesh National Book Centre (2003).
* The exact number of special libraries is not known, as this has not been investigated up to now.

frustrating. Though college libraries, especially a private college library based in Dhaka offers internet services to their students and faculty members, use of ICT in most college libraries is very restricted. Almost all special libraries based in Dhaka offer internet service to their users. Special libraries outside Dhaka offer limited internet services. During this study the author talked to many library professionals in Bangladesh to find out the exact potential for m-library service development and the attitudes of the respective libraries and librarians towards new service development. Survey results showed that it would be possible to implement m-library services in all public and private universities in Bangladesh, and in most special libraries based in Dhaka. Unfortunately, the current status of ICT in other types of libraries is far from m-library service development.

Mobile companies in Bangladesh

Mobile phones came to Bangladesh during the early 1990s. At that time City Cell offered mobile services to a limited number of people. From 1997 the sector was opened up to some other companies and a revolution started in the mobile sector. However, it was only from 2000 that the number of mobile subscribers increased substantially. Bangladesh has currently six mobile phone operators. Table 24.2 (on page 226) shows these companies and their numbers of current subscribers:

The total shows that 47.97 million mobile users could also be library users. Before discussing how, we will briefly examine the information services provided by Bangladesh's six mobile phone companies.

Grameenphone (GP): The leading telecommunications service provider in the country; offers the following information services using mobile technology:

News service: Latest news from Channel 1, CNN, bdnews24.com, etc.

Weather: Weather update, on request from users for weather update service.

Stock information: Live information on stock price details; daily graph of prices of trading companies; alerts on stock price changes on the Dhaka and Chittagong Stock Exchanges via Bull. Bull is a Java application for mobile phones which displays stock information of Dhaka and Chittagong Stock Exchanges using EDGE/GPRS.

Healthline: Medical advice on emergency, non-emergency or regular medical situation by dialling 789. This service received GSMA Award 2007 for 'Best use of Mobile for Social and Economic Development'.

Banglalink: Offers the following information services using mobile technology:

Banglalink jigyasha 7676 (Ask Banglalink): Suggestions and answers in response to queries related to agriculture, vegetable and fruit farming, poultry, etc., by dialling 7676.

Healthlink service: Health counselling services, by dialling 789.

Railway information service: Railway information service for users from Banglalink Junction 1313.

Yellow Pages: Value added service enabling users to access Bangladesh Yellow Pages via Banglalink mobile phone.

i'info from Banglalink: SMS-based subscription news service (see www.banglalinkgsm.com/html1/iinfo.php for further information).

Aktel: Offers the following information services using mobile technology:

Aktel directory service: Access to information, anything from household needs to life-saving information.

Aktel Cafe8000: Voice based services such as news, entertainment, international roaming information, etc.

News service: Access to latest news by typing 'news' and by SMS to 2324, or call 2324 anytime.

CityCell: Offers the following information services using mobile technology:

News: 24-hour access to leading national news from different news media by dialling 2010, 2626, 4141, etc.

Train schedule: Access to train schedules by dialling *125.

Exchange rate: Current foreign exchange rates by dialling *126.

Teletalk: Offers the following information services using mobile technology:

Push-pull services: Service for important information such as cricket updates, weather forecasts, prayer times, quotations, horoscopes and especially *sehri-iftar* (sunrise/sunset) times during Ramadan, etc.

Warid Telecom: Information services via the Warid SIM Genie menu – the next generation SIM Tool Kits (STK)! The SIM Genie menu is presented as soon as the Warid phone is switched on. Information services are provided under different categories, including:

Info zone: Daily sport, news, finance, religion and weather information services, dictionary, etc.

Entertainment: Download service for logos and icons, jokes, ringtones, music and movies, quotations, horoscopes, games and recipes.

Help and support: Helpline for direct help from Warid Contact Centers, emergency services such as police, fire and ambulance by one-touch dialling.

Table 24.2 Number of mobile operators with their corresponding subscribers

Operator	Year started*	Active subscribers (million) (July 2009)**
Grameen Phone Ltd. (GP)2	1997	21.15
TMIB (Aktel)3	1997	9.90
Orascom Telecom Bangladesh Limited (Banglalink)4	2005	11.27
PBTL (City Cell)5	1993	1.96
Teletalk Bangladesh Ltd. (Teletalk)6	2004	1.08
Warid Telecom International L.L.C. (Warid)7	2007	2.61
Total		47.97

Sources: * Compiled from the official websites of the mobile operators.

** Bangladesh Telecommunication Regulatory Commission (2009)

Possible library services using mobile technology

How could 47.97 million mobile users become library users? Quite simply, by offering library services via mobile technology. Library services that could be offered via mobile technology are briefly discussed below.

Current awareness service

Any library could offer a subscription-based current awareness service via mobile technology. Subscribers could receive announcements, library updates, etc., and monthly subscription fees could be automatically deducted from the user's mobile account. Libraries would negotiate with content providers for service support. For example, to subscribe to the CAS (current awareness service) a user would need to type 'On CAS' and send the message to the 5000 number from their mobile, as shown in Figure 24.1.

After sending a message to 5000, the user would receive a confirmation message on their mobile, as shown in Figure 24.2.

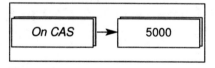

Figure 24.1
Sample CAS service subscription method

> Congratulations. You have been subscribed to CAS. To unsubscribe write Off CAS and send to 5000. 15 taka will be deducted from your mobile balance for this service for one month.

Figure 24.2
Sample reply from the server after user subscribed to CAS service

Book renewal service

A user who is on the move or unable to visit the library to return a book on time could use a book renewal service. For example, the user could type the ISBN number of the book with their library membership ID and send the information to BOOK (2665). The library server would then send a confirmation message to the user with the new book return date, as shown in Figure 24.3.

> Your book entitled 'Information Seeking in the Online Age' by Large, Andrew et al. is renewed for 21 days. It will expire on July 15, 2009.

Figure 24.3
Sample confirmation message from the server after user sent message to 2665

Book alert service

This request-based service would provide information on new acquisitions in different subject areas. For example, an Information Science faculty member or student wishes to know what new acquisitions the library has in their subject. The user types in a subject code, as assigned by the library, and sends it to 1111. The server replies by reporting the number of new books/journals on that subject that it has received during the last 30 days (Figure 24.4). In order to provide this service a library would need to assign subject codes to its holdings.

We have 05 new books and 01 new journal issue on your subject. Please visit the library.

Figure 24.4
Sample message from book alert service

Overdue books notice

You have no overdue books/journals.

Users can use mobile technology to receive overdue book notices. They can request such a report by sending their library user ID to 1234 (Figure 24.5).

Figure 24.5
Sample message from the server after sending message to 1234

Disaster information service

Every year Bangladesh experiences a variety of natural disasters which take thousands of people's lives, especially in the coastal area. By providing weather updates to the fishermen and sailors before the weather worsens, public libraries in the coastal area could help to save the lives of many people. In order to provide this service the public library authority would need to have a database register of fishermen and sailors, with their mobile numbers. As the majority of fishermen and sailors are illiterate, the service would need to be provided by voice message. The library could even provide voice-based live support in the event of a disaster, to help people to find shelter or relief.

National health information service

Health information services could be provided centrally to all mobile users by the Information Centre of the Ministry of Health. The service would include both individual health advice from professionals, and general health support during an emergency. The service would be an interactive one, accessible to anyone at any time. A health information directory would enable the service to provide information on locally available healthcare, for example, the name and address of a cardiac specialist in Kishoregonj district. Other information available would include primary healthcare tips, beauty tips, etc.

Agricultural information service

An agricultural information service could be a very effective service for farmers and for people involved in agricultural business. The local Agricultural Information office of each division could start this service using mobile technology. The Agriculture Ministry of the Peoples' Republic of Bangladesh should centrally coordinate this service. As most farmers and local people involved in agricultural business are not literate, a live information service would be needed. The user would phone the local agricultural information office to request information or advice, e.g. the current market price for rice, or the type of crop that is suitable for a particular field, etc. The operator would either answer the question immediately or find out the answer, if necessary calling back to the user who made the request for information.

One-stop services of the Government of Bangladesh

The Government of Bangladesh has recently introduced a Right to Information Act, ensuring people's right to know government and other information whenever required. A 'One-Stop Information Centre' could be established to comply with this law. The centre would provide all kinds of information from the various government ministries. Examples of questions that could be asked via telephone are: Who is the Secretary of a particular ministry? What projects has a particular ministry undertaken recently? How can the various officials of the secretariat be contacted? And so on.

Other information services using mobile technology

Other possible information services that could be provided via mobile technology include the following.

Entertainment service: Different types of songs, short dramas, recitations, etc. could be stored on the public library's server. By dialling 6666 users would be connected to the library's entertainment server and would be able to choose songs, poems or dramas to listen to. This should be a free service.

News service: Besides providing printed newspapers on premises, public libraries could also provide news headlines and a news briefing service via mobile technology. By dialling 7777 a user would enter the public library's news room and be able to hear the latest news. Libraries could enter into an agreement with the different news media to provide this service. Such a service is already available from the different mobile operators, but the charges are rather high. The public libraries could persuade the mobile operators and the news media to provide the news service as part of their corporate social responsibility.

Traffic updates: Bangladesh's capital city has serious traffic jams throughout the day. Traffic updates could help people choose the best route to their destination. This service could be provided by a 'Metropolitan Traffic Department Information Centre'. Information professionals in the department would give voice updates on various important roads in Dhaka every half an hour and users could receive the latest traffic update by dialling 1000.

WAP content development: A good number of mobile users in Bangladesh use WAP-enabled handsets, and the number of mobile internet users is increasing all the time. Availability of some library content via a WAP site would undoubtedly increase library access. A particular library's WAP site might include the following facilities:

- library catalogue
- library opening hours
- library contact information
- some external links for mobile content
- directory information for the particular library, ministry or university
- Yellow Pages.

How to implement m-library services

Discussions with providers of various mobile services and developers of mobile content indicated that implementation of mobile services required a database of information, 24-hour server support and support from a content provider. If the library has no server, the content provider should be able to supply one. In the initial stages this could be costly, and costs would vary from one content provider to another and depending on agreements with mobile operators. If a library could convince the mobile operators that their service fulfils a social need, for example a health service, it is possible that operators might not charge for it.

All these services would be possible only if the government and concerned authorities, as well as the mobile operators, came forward to implement them. The government would need to take the steps necessary to involve the mobile operators and to improve library services using mobile technology. Without support from the mobile operators, none of the proposed services could be implemented.

Conclusion

Bangladesh has a greater number of mobile subscribers than many other countries in the world, and there is a great potential for development of mobile services. Mobile services would be quite easy to develop, and the government could initiate mobile information services immediately under the Right to Information Act of Bangladesh. Such a service would be particularly valuable during crisis periods, such as a public health emergency. At such a time the government could send out health warnings by text message to all mobile subscribers; interactive voice-based information services should also be available during any national crisis. By dialling a designated crisis information number, e.g., 111 users could be connected to the crisis information service and receive information updates.

This chapter has aimed to illustrate the basic services that can be offered via mobile technology. All the services mentioned are model services for mobile library content/service development worldwide. A revolution would take place in the field of library and information science if libraries and information centres around the world acted now to give mobile library services a 'kick-start'.

Notes

1   National Web Portal of Bangladesh,
www.bangladesh.gov.bd.
2   Grameen Phone,
www.grameenphone.com.
3   Aktel,
www.aktel.com.
4   Banglalink,
www.banglalink.com.
5   Citycell,
www.citycell.com.
6. Teletalk,
www.teletalk.com.bd.
7   Warid,
www.waridtel.com.bd.

References

Bangladesh Association of Software and Information Services (BASIS) (2005) *Study on the IT sector of Bangladesh*, BASIS.
Bangladesh National Book Centre (2003) *Library Directory {Bangla} Dhaka*, Bangladesh National Book Centre.
Bangladesh Telecommunication Regulatory Commission (BTRC) (2009) *Mobile phone subscribers in Bangladesh*,
www.btrc.gov.bd/newsandevents/mobile_phone_subscribers/mobile_phone_subscribers_july_2009.php.
Department of Public Libraries (Bangladesh) (2009) *Public libraries at a glance*,
www.publiclibrary.gov.bd/index.php?option=com_content&task=view&id=382&Itemid=427.
Khan, M. S. I. (1989) Developments in New Information Technologies and their Applications and Prospects in Bangladesh, *Media Asia*, 16 (1), 32–40.
National University of Bangladesh (2009) *Division Wise Affiliated College List*,
www.nu.edu.bd/statistics/division_wise_college.pdf.
Shuva, N. Z. (2005) Implementing Information and Communication Technology in Public Libraries of Bangladesh, *International Information & Library Review*, 37 (3),159–68.
University Grants Commission of Bangladesh (2009a) *List of Public Universities*,
http://ugc.gov.bd/university/?action=public.

University Grants Commission of Bangladesh (2009b) *List of Private Universities*, http://ugc.gov.bd/university/?action=private.

25

M-libraries: information use on the move

Keren Mills

Introduction

At the First International M-Libraries Conference in 2007 it was clear that many libraries were keen to explore the potential of mobile library services, but none of the delegates reported having asked library users what sort of library services they might want to access from mobile devices, or which mobile devices they were comfortable using. When I had the opportunity to take part in the Arcadia Programme at Cambridge University Library, I opted to investigate how library users access and interact with information when they are on the move. The key questions posed to students were 'Do you use mobile devices?' 'What do you use them for?' and 'What would you want to access from the library?'

Method

I surveyed staff and students studying at both the Open University (OU), a distance learning institution, and Cambridge University, a collegiate institution. The survey focused on their use of information and, more specifically, on how they accessed information or used it on their mobile phones.

With only ten weeks in which to complete the research and approaching the end of term at Cambridge, I chose to use an online survey to gather data. The survey was promoted via mailing lists and library websites with the incentive of being entered in a prize draw. Only in the case of Open University students did the survey go to a selected sample of 2000

Table 25.1 Survey response rates (percentages)

Survey group	Cambridge University	Open University
Undergraduate students	49%	46.6%
Postgraduate students	25%	10.4%
Associate lecturers	–	21.4%
Academic staff	9%	9.0%
Academic-related staff	4%	10.6%
Secretarial and clerical/assistant staff	7%	1.9%
Total number of respondents	1530	776

students, broadly representative of the student population. All other respondent groups were self-selecting.

Realizing that the provision of library services to mobile devices is a new area and that I didn't have any sample services to demonstrate to library users, I thought it likely that survey respondents would be confused or put off by questions relating specifically to provision of library services, and so I decided to focus instead on how they used their mobile phones to access and interact with information. Table 25.1 shows the total number of survey participants, and the response rates for the different user groups identified.

Results

Survey respondents were more positive about accessing information via text message than through the mobile internet. This can be attributed to both ease of use and the perceived costs of the two methods.

Over 60% of survey respondents had owned their current mobile phone for less than two years, which indicates that many of them had fairly up-to-date devices, with functionality including basic mobile internet browsing. With such a high percentage of the academic population upgrading their phones every one to two years, it will be important for libraries offering m-library services to track trends in new functionality and new standards for mobile phone handsets. For instance, many phones released in 2009 emulate the iPhone's touch-screen interface and are likely to try to compete with the improved internet browsing functionality that it offers.

The majority of respondents used their phones primarily to make calls, send text messages and take photographs, although they liked to know that the other functions were potentially available. Respondents' use of different forms of media on their mobile phones was mostly limited to viewing photographs. Some used their phones to listen to music or watch videos, but very few used them to listen to podcasts or audio books, and only a small number read e-books or e-journal articles. Some commented that they prefer to use their iPods or other media players to access these other forms of media.

I discovered that although a variety of mobile devices are available, mobile phones arguably require the most specialized development in terms of delivering appropriate presentations of information services.

SMS alerts/notification

A number of banks, transport services and other services offer SMS (Short Messaging Service) alerting services. London residents can sign up for text alerts from Transport for London to let them know the status of underground lines on their route. Roughly 32% of all respondents had signed up at some time for SMS alerts of this nature, and 34% of these still received them. Two separate pilot studies undertaken at the Open University (Carberry, 2008) and the University of Wolverhampton (Brett, 2008) found that students liked receiving SMS alerts from the university, provided that they were not too frequent. These studies also found that, given the 160-character limit for text alerts, they have to be carefully phrased to ensure that they are easily understood and do not sound too abrupt.

The enthusiasm for SMS alerts was greater at the OU than at Cambridge, but at both institutions a significant proportion of respondents were using text alerting services in some form. I found that students would be in favour of receiving text alerts from the library to let them know when reserved items were ready for collection and when books were due or overdue for renewal/return. Overall, 21% of all Cambridge respondents were in favour of text alerts from the library, as compared to 35% of all respondents at the OU. Comments from respondents indicated that many would like to receive these notifications by both text message and e-mail.

SMS reference

The term 'SMS Reference' is used here to refer to services that allow the user to send a query by text message and receive a reply the same way. A popular example of this is Any Questions Answered (AQA 63336). When asked whether they had ever used such a service, 27% of respondents said they had, and another 26% said they might try it now that they were aware of it. Only 4% of those who had tried it would not use such a service again. From these figures we can deduce that it might be worth piloting a service allowing users to submit queries to the helpdesk via text message. Certainly, many libraries that have implemented 'chat reference', allowing users to have a live conversation with a librarian through instant messaging, have found it to be very popular (Hvass and Myer, 2008). As with text alerting services, these enquiries could be received and dealt with by staff through an e-mail or web-based interface; this would reduce the amount of time taken to type a reply, as compared to using a mobile phone keyboard. Other presenters at M-Libraries 2009 had piloted SMS reference services and found them to be quite popular with their users (see Chapter 26, this volume).

Mobile OPAC

Staff at Cambridge University Library have observed patrons using their camera phones to take pictures of the catalogue results screen, rather than noting class marks on a piece of paper. Fifty percent of respondents at both universities said that they took photos of signs, books, etc. to save information for later use. In addition, 55% of total respondents were in favour of being able to access the library catalogue via a mobile phone. In the short term, libraries could allow patrons to use their phones for notes and photos within the library, so long as they are on silent or in flight mode. In the long term, libraries could work with their Library Management System (LMS) supplier to create a mobile version of their library catalogue. There is already a mobile application for OCLC's WorldCat, so libraries who submit their catalogue records to WorldCat could make use of this application to pilot the service.

Mobile content delivery

Respondents were asked whether they used their mobile phones for any of the activities listed in Table 25.2.

Table 25.2 Media use on mobile phones		
Activity	Percentage of respondents who never do this	
	Cambridge University	Open University
Read an e-book	93.8%	92.3%
Read e-journal articles	91.5%	86.4%
Listen to podcasts or audio books	87.5%	78.1%
Listen to music	60.0%	56.7%
View photos	37.8%	32.3%
Watch videos	69.4%	61.5%

The results suggest that rather than libraries putting development resource into delivering content such as e-books and e-journals to mobile devices, it would be more cost-effective to encourage content providers to do the development work. This would also address licensing issues, as many subscription licences to academic e-texts prohibit downloading of the content.

Some mobile phone devices, such as iPhones and Windows Mobile devices, can already display e-books. Audio files such as podcasts and audio books can easily be played on many mobile phones or portable media players. At the time of the survey, however, most users were put off by the constraints of the technology, such as poor screen quality. However, comments from the survey respondents suggested that iPhone users were already more inclined to read e-books on their phones. Athabasca University has developed a Digital Reading Room to enable its users to access e-journals, even when the publishers do not support mobile access to their content.

Mobile internet

The number of smart phones available is increasing, but many mobile owners still restrict use of their phones to calling and texting. The iPhone and Android phones have led to an increase in mobile internet use in the UK, and as other mobile phone manufacturers release competitive devices this is likely to increase further. However, the key difference between these and previous web-enabled mobile phones is that the iPhone and Android models can comfortably access websites intended for larger screens. As this type of device becomes increasingly available it will no longer be necessary to develop mobile-ready websites.

According to a report from Continental Research, the perception of many people in the UK is that the mobile internet is expensive, slow and difficult to use. The report also states: 'Internet access has improved significantly in recent years, but the proportion that has used the service has remained stubbornly static and is unchanged over the last year, at 12% of mobile owners' (Continental Research, 2008).

Several survey respondents commented that they might access the internet if they had access to a larger-screen device, such as the iPhone, and others commented that they prefer to use their laptop or netbook.

Less than 16% of Cambridge respondents used their mobile phones to access the internet more than once a week, and only 25% at the OU. While this was slightly higher than the national average at the time, one obvious inference to be drawn is that it is not worth libraries putting time and effort into developing dedicated mobile websites. If libraries want their sites to be mobile friendly, it would be more advisable to use either Cascading Style Sheets (CSS) or Auto-Detect and Reformat software (ADR) to facilitate rearrangement of content and navigation to suit the sizes of mobile screens. This would ensure both a sustainable approach and that library websites presented content well on all sizes of screen, including the popular netbooks. The survey results showed that what people were most likely to look up on the move were the library's opening hours, contact information, location and OPAC, and the user's borrowing record (see Figure 25.1). There could be some value in ensuring those features of the website were only one click from the home page. (The survey question about location for Open University respondents was whether they would like mobile access to a map showing the nearest library they could use, since relatively few Open University students are able to visit the university's own library building in Milton Keynes. Their access to library resources is primarily electronic, but through the SCONUL Access[1] scheme they are able to use other university libraries as study spaces or to borrow books.)

Library applications (software) for mobile phones

For some years, people have been creating applications that can be downloaded to mobile phones in addition to the software supplied with them. Examples include mapping software, games and mobile e-mail clients. Until recently the take-up of these applications was relatively low,

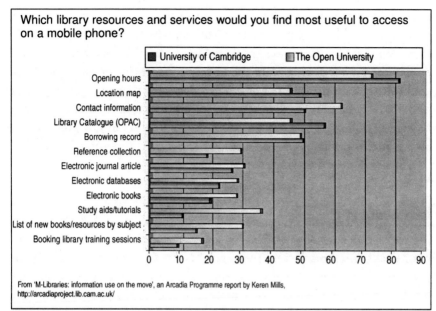

Which library resources and services would you find most useful to access on a mobile phone?

From 'M-Libraries: information use on the move', an Arcadia Programme report by Keren Mills, http://arcadiaproject.lib.cam.ac.uk/

Figure 25.1 Mobile access to library services

but iPhone users download a great many more applications than do owners of other phones (ComScore, 2009), and owners of Android phones are likely to follow suit.

Only 21% of respondents in this survey had downloaded applications to their phones and would do so again. With such low download rates there seems to be little current value in providing library applications. It would be more cost-effective to provide the same functionality through a website.

Library audio tours

A quarter of respondents would like access to audio tours of the libraries. Responses indicated that only 9% would download audio tours to their mobile phones, but 16% would download them to their MP3 players and 19% would like to borrow an MP3 player pre-loaded with audio tours from the library counter.

Audio tours can be produced fairly quickly and inexpensively, so libraries that run inductions throughout the year or have a poor attendance rate at induction sessions for new students, might find that tours would reduce the amount of staff time spent helping new users to orient themselves in library environments and explaining the facilities. Audio tours could easily be provided as downloads from the library website and on devices for borrowing from the library counter. In the long run, they could also reduce the number of staff-guided tours that some library users find distracting. One disadvantage of some audio tours is that they often have to be followed in a set, linear order. Some users might prefer more choice and flexibility, and even a simple paper map and guide in addition to a location-aware mobile tour.

Conclusions

Mobile phones are still viewed by the majority of users as devices for making phone calls and sending text messages. Users often don't associate them with other activities, such as information seeking. However, users are increasingly dependent on their mobile phones and a growing number do use them as diaries, for taking notes and for e-mail and internet access. The indications of this study show that there is likely to be an increased expectation among library users that libraries will provide some services in a mobile-friendly way.

It would be beneficial to build on the work done in this study and to undertake further piloting to identify where development could add value, e.g. allow users to respond to text alerts using single words or short phrases to act on those alerts, as in the examples shown in Table 25.3. This would enable users to respond to an alert immediately, rather than having to remember to act on it later. The functionality of mobile phones and the

Table 25.3 Text alerts and responses		
Alert	**User response**	**Library action**
[Item title] is due for renewal in 1 day. Reply with 'RENEW' to keep this item or 'RENEW ALL' for all loans.	RENEW RENEW ALL	Renews named item Renews all loans
The item you reserved is ready for collection. [Item title]	CANCEL	Cancels reservation as user no longer needs that item.
	STOP	Stops sending notifications as text messages, reverting to e-mail.

availability of services from network providers will continue to change, so demand from users for m-library services and choices may well increase.

Recommendations

These recommendations are based on the combined responses from both Cambridge and Open University survey participants, and indicate that higher education libraries should consider the following:

- Piloting SMS alerting services, giving users the opportunity to choose whether to receive notifications by text message, e-mail or both. Such services are likely to be taken up by at least a third of library users. Alerts would include notifications automatically generated by the LMS.
- Piloting a SMS reference service, if the library receives a high volume of enquiries that require brief responses, such as dictionary definitions, facts or service information.
- Providing a mobile OPAC interface, possibly using a service such as AirPAC or WorldCat Mobile, or working with the library's LMS supplier to develop a mobile version of the OPAC.
- Ensuring that the library website is 'platform agnostic' and accessible and will resize to smaller screens, in order to position services to respond to increasing numbers of netbook users and mobile internet users.
- Allowing mobile phone use in the library, as long as devices are set to silent or to flight mode (meaning they are not receiving a signal) so as not to disturb other users.

To mobilize the Open University library, staff have developed a mobile-friendly website[2] and mobile revision activities for an online information literacy tutorial, Safari.[3] The Systems Development team is working to enhance the search functionality available via the mobile site (Chapter 9, this volume).

Notes

1   SCONUL Access (2009),
    www.access.sconul.ac.uk/

2   www.open.ac.uk/library
3   http://digilab.open.ac.uk/testarea/mobileSafari/

References

Brett, P. (2008) *Mobiles Enhancing Learning and Support (MeLAS)*, MeLAS,
www.jisc.ac.uk/whatwedo/programmes/elearninginnovation/
melas.aspx.

Carberry, G. (2008) *OU Texts/SMS services*, The Open University,
http://kn.open.ac.uk/workspace.cfm?wpid=8516.

Continental Research (2008) *Autumn 2008 Mobile Phone Report*, MRC0809/24B.

ComScore (2009) ComScore releases first data on iPhone users in the U.K. Abstract.
2009, no. 30/03/2009

Hvass, A. and Myer, S. (2008) Can I Help You? Implementing an IM service, *The
Electronic Library*, **26** (4), 530–44,
www.emeraldinsight.com/10.1108/02640470810893774.

26

UCLA and Yale Science Libraries data on cyberlearning and reference services via mobile devices

Brena Smith, Michelle Jacobs, Joseph Murphy and Alison Armstrong

The role of SMS in information services

Mobile phones are quickly emerging as a primary information interface for many people. Mobile devices are becoming as important as desktops, and even as important as laptops, as tools for engaging with information. Text messaging (SMS) is a centrepiece of mobile communication worldwide because it is universal across phone types, allowing it to bridge the smart phone/basic mobile phone divide.

SMS (Short Message Service) is used for so much more than communication: it's now a major medium for seeking, applying and transferring content. We seek information from our friends and from the crowd with online social networks like Twitter and Facebook. We also use SMS for knowledge management activities such as tracking and sharing information and resources. SMS is thus a central part of how we engage with information, and libraries can stay relevant and maintain their role as information centres in the evolving mobile environment by offering services to their patrons using text messaging reference.

Technology options for SMS reference

Choosing the most appropriate technology for SMS reference in your library is a major concern affecting all areas of the service, especially management and staffing, with implications for the service's long-term success. The major considerations in evaluating technologies and products are: cost, ability to meet patrons' expectations, sustainable staffing workflows, interoperability and flexibility. Beware of tools that require additional steps for the patron or the librarian. The best technologies allow patrons to send messages directly to a normal phone number with no additional steps and allow librarians to respond directly, in one step.

The major technology options for SMS reference are free SMS to instant message (IM) mash-ups, basic mobile phones or smart phones, web-based tools for receiving text messages in a web platform, SMS/e-mail tools that convert text messages to e-mail, and workarounds such as Twitter or Google Voice. Each technology has its strengths and weaknesses. The SMS/IM (instant messaging) options, like the web-based options, require patrons to enter a screen name in the message in addition to the short-code address. Both the SMS/IM and the SMS/e-mail options have the benefit of incorporating the service into existing virtual reference programs, but separate the librarian from the technology used by patrons.

Using a mobile phone to receive and reply to text messages offers the most flexibility, giving librarians the same mobility as their patrons and forcing them to use the same communication norms (abbreviations and character limits) as patrons. Using a mobile phone entails equipment and service fees, and demands new staffing structures.

It is difficult to gauge librarians' and users' adaptability to the changing foci of information-seeking behaviour and technology habits. The future applicability and flexibility of the technology should be weighed carefully, because a product that cannot evolve with and adapt to emerging technologies will not be useful for long. For instance, mobile phones can provide reference via Twitter and Facebook, while web-based tools may not be able to interact with the mobiles at all. It is the responsibility of librarians to have first-hand familiarity with mobile technologies so that they can make informed decisions when selecting tools.

Advanced management concerns for SMS reference

Developing SMS reference services involves a new set of considerations, which come with meeting the evolving world of mobile information. Staffing and supporting mobile technologies requires a shift of focus from static services and places to dynamic mobile spaces, which bring new challenges in workflow and training. From a management perspective, how the service is staffed and scheduled may need to change, in order to ensure that the quality of excellence of the traditional reference desk is also applied to mobile services. This can be challenging if staff do not have experience with mobile devices and using text message shorthand.

Developing a proposal for text reference

It can take time for a new service to become successful and library administrators and managers may not see advantages immediately. In tight budget times, services without a quick uptake may not be carried forward. However, our patrons are mobile, information is mobile, and if librarians stay behind the desk, we will be at a disadvantage in the mobile landscape. SMS is a transformational technology that plays a role in mobile work platforms including those for internal communication through mobiles and with social networking sites. The major management considerations for this landscape are user expectations and training for staff.

It was for these reasons that we sought to offer text reference services. Further, it offered students another option for communicating with reference librarians, thus increasing our suite of reference services to include face-to-face, e-mail, IM chat and text services. The project was approved as a one-year test.

Choosing the service and working with the vendor at UCLA

Finding the best way to provide the service was of the utmost importance, as a balance was needed between the librarians' needs for multiple devices to provide the service and the users' needs to be able to send text messages as usual. At UC Merced, where Michelle Jacobs had already started a text reference service, it was simpler to offer the service because the librarians used mobiles as their primary phones, thus allowing students to text specific librarians. UC Merced had an FTE (full time equivalent) of approximately 2000 students at the time and the model was feasible. Adapting it for an institution the size of UCLA, with an undergraduate FTE of approximately 26,000 and where librarians did not have mobile phones required a different back-end model. Initially the library had hoped to purchase a mobile device that would be shared with the librarians staffing the service. This in itself created some problems, but the largest issue was the change in tax regulations in California which made it difficult to purchase mobile devices in State institutions. This led us to search for a virtual SMS provider.

There were few options of virtual SMS provider at the time. Upside Wireless, the vendor ultimately selected, was willing to work with UCLA and develop tools specific to its needs. We were able to integrate with MS Outlook, making it easier for staff to respond to incoming SMS. Upside Wireless also had a competitive price structure and worked very closely with staff to develop the product over the next year. The virtual SMS is now seamlessly integrated into e-mail and librarians can respond to an incoming SMS, just as easily as they would to any other e-mail, but with a limited number of characters, of course.

Choosing a service at Yale

After careful investigation, Yale Science Libraries chose to use an Apple iPhone because of its flexibility and robust features. We use the iPhone to cover all reference services: IM, e-mail, in person and Twitter. We were able to purchase the devices using a budget fund specifically designated for technology purchases.

Alignment with the UCLA Library strategic plan

An important component of proposing any new programme or service within the UCLA Library is evaluating how the service aligns with the strategic plan. This helps to ensure that we are supporting the mission of the library and working in conjunction with the other library units across the campus.

The two strategic goals that we focused on were 'enriching services' and 'improving research skills' (UCLA Library Strategic Plan, 2005, 11–13). SMS reference enriched services in two ways. First, by implementing SMS, we added another direct service point to a reference librarian. Second, we added a dimension to our service point, in that students could access from *any* physical location. This is a benefit to both students and librarians: not only can students be on the go, but the librarians can be, too.

SMS reference addressed enhancement of research skills because it allowed librarians to deliver academic support by a means that was already used by students in their personal lives. While we were unable to gather statistical information on this point, based on the queries we received from students, we believe that some of our text users are students who might not otherwise seek help from a librarian. SMS reference offers a 'safe' way for students to ask for help, especially for those students who may experience 'library anxiety'.

Implementation at UCLA

Implementation of SMS reference at UCLA was carried out through a pilot project during the 2008–9 academic year. Prior to implementation, we had targeted two libraries on campus in which to roll out the service and three librarians to staff it: the Louise M. Darling Biomedical Library, which serves the health and life sciences disciplines, and the College (Undergraduate) Library, which serves the undergraduate students. Two librarians from College Library and one librarian from the Biomedical Library were the staff to launch the service.

Unsure of how the service would be received and unwilling to be victims of sudden success, we chose a soft launch of SMS reference for fall 2008. We did very little marketing and promotion for the service, beyond distributing fliers and discussing the service during instruction sessions and other events that were attended by librarians. A soft launch allowed us to work through staffing and technology issues and to see how the service operated.

During winter 2009 we ramped up our promotional efforts. We tasked interns from the UCLA Graduate School of Education and Information Studies, who were working in the College Library, with creating and carrying out a marketing plan. Working with UCLA Library's Director of Communication, the interns were sure to adhere to our branding policies and were able to brainstorm various possibilities for promoting the SMS reference service. Ideas from brainstorming included:

- creating a short video
- buying an ad in the student newspaper, *The Daily Bruin*
- identifying offices, departments and various student groups (academic and non-academic) for distribution of fliers and to talk with office employees
- modifying the flier that had been created the previous fall
- creating fliers in multiple languages for the international student centre.

Whittling down the brainstorming list because they had multiple tasks as part of their internship, the interns selected fliers as their focus. Why use SMS reference? This question was faced continuously when discussing the service with students and librarians. There was confusion about how, why and when this service could and should be used. To help curtail the confusion we looked at the questions we had received during the fall and selected about 20 as representative of the types of questions that came in to SMS reference. We put these questions on the back of our flier to illustrate to students and librarians the best practice for the service. They proved to be an excellent addition to the fliers, and helped enormously when we marketed the service to students, and internally when communicating to other librarians and our senior management team.

The interns created fliers in multiple languages. Although we do not support SMS reference in other languages, the fliers were used as a promotional material for distribution in the Dashew Center for International Students. Throughout the winter, the interns distributed batches of fliers to offices and academic departments that were frequented by students. They took the opportunity to discuss the services with people working in the offices. These actions, together with some small changes made to the fliers and the co-ordinated flyer drops throughout the campus, had a significant impact on our statistics. We saw a dramatic increase in usage.

Implementation at Yale

We planned policies to meet service considerations, then tested the technology and workflows, trained staff, refined staffing models and marketed to staff and users. We continually re-evaluate practices and policies.

Staffing the service at UCLA

Staffing for SMS reference was voluntary throughout the pilot. During fall 2008, the service was staffed by three librarians – two from College Library and one from the Biomedical Library. The schedule was managed by an internal, web-based calendar, and matched the regular reference desk hours of the College Library: Monday through Thursday, 9 a.m. to 8 p.m.; Friday, 9 a.m. to 6 p.m.; Saturday and Sunday 1 p.m. to 5 p.m. Because there were a large number of hours of service and only three librarians to staff them, the service shifts were divided into time blocks of four to five hours each. The long shifts were not a problem because our service provider, Upside Wireless, allows us to receive incoming messages to both phones and computers, thus enabling librarians to provide the service from any location.

As service usage increased during winter, we trained the interns to begin staffing the SMS and put out a call to others within College Library to help with service provision. We trained three additional people – one librarian and two library assistants. We also restructured the shifts to one-hour shifts and began staffing SMS while serving during our regular reference desk shifts.

Staffing the service at Yale

Our staffing model balances our diffuse nature (six science libraries) with the chosen technology and perceived user expectations. We share one iPhone amongst librarians in the Central Science Library and offer coverage on the same schedule as reference desk shifts. We engage in ongoing training for regular updates to the phone software and as refreshers.

How text reference through virtual SMS works

The information flow is illustrated in Figure 26.1.

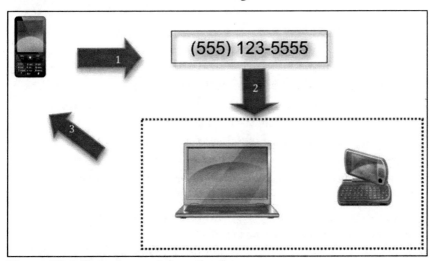

Figure 26.1 Text reference information flow

It is as follows:

1 The question comes through the mobile device to a standard 10-digit phone number.
2 The message is then sent to assigned devices. These can include: any number of e-mail accounts and other mobile devices. The question is then replied to through either module.
3 The user receives the response via text message.

Best practices

In order to establish a good model of service, UCLA developed a set of best practices, which were based on best practices followed for both digital reference and traditional reference desk service. These include:

1 Limit your response to 140 characters. It is OK to use 'text', such as: 'If u hve more? txt back l8tr'. This practice was modified during the course of the year. Text messages have evolved to more of a chat-like communication and our model was adjusted to match this trend.
2 This form of communication is commonly informal and, due to the limited number of characters, you do not need to introduce yourself.
3 If it is going to take some time to research the answer, send a text letting the user know when you can respond.
4 If the answer is lengthy, offer to reply by e-mail.
5 Some questions may require referrals. Try to refer the user to a person, not just a department, and provide the contact information.
6 Do not call the patron. If a conversation is needed, text to see if you can call them or if they would prefer to call you.
7 If you are helping with research and would like to send the patron some URLs or articles, ask for their university e-mail account. This allows the library to verify that the user is part of the university community before the information is sent.
8 If you are not sure how to respond to a request, ask the advice of a member of the text reference staff.
9 If the question is inappropriate, do not engage with the patron by responding.

SMS reference statistics at UCLA

The most common types of question that we receive are what librarians refer to as 'ready reference' queries (Figure 26.2). In general, these questions are not much different from the types of question received face-to-face at the reference desk. Examples include students seeking information about the availability of a book or items in reserves, or quick, factual information on any given topic. As to questions about off-campus access, we received so many that we treated them separately in our statistics. We also receive directional questions regarding the location of particular buildings and, interestingly, we received a number of questions from non-UCLA users, usually regarding their ability to access library materials.

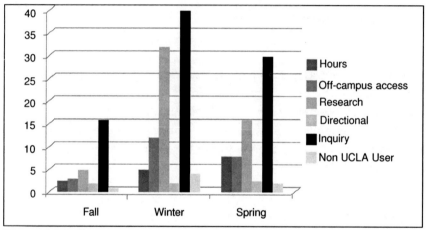

Figure 26.2 Query types

As expected, we had a slow start in fall 2008, when we launched the service (Figure 26.3), receiving a total of 28 queries, most of them between the sixth and ninth weeks of the quarter. We saw a large jump in usage during winter, with a total of 93 transactions. As stated earlier, we attributed this to the increase in marketing activities during the quarter. We believe that word of mouth amongst students also help increase usage of the service. Similar to the fall, the weeks with the highest usage in winter and spring were between the fifth and ninth weeks.

During all three quarters there was steady usage in the first two weeks of the quarter. Many of the questions we received were directional or about specific library services, and they were typical questions for the start of

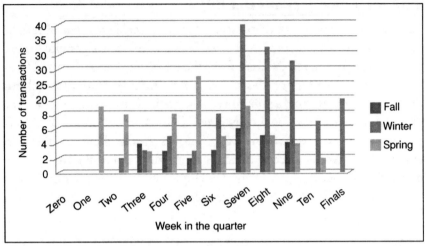

Figure 26.3 Transactions

a quarter. There was usually a dip in usage during weeks three and four. The service would then pick up again in weeks five or six and then climb steadily until week nine or ten. We believe this reflects how students engage in research throughout the quarter, i.e. working on general course work early in the quarter and then working on research assignments later in the quarter.

Conclusion: the future of SMS reference

The future of SMS reference lies in closing the gap between patrons' expectations and practical staffing limitations. We expect libraries to turn to consortia for staffing and product support in the short term, and then a return to mobile devices as libraries embrace the tide of mobile technology. We ultimately envision libraries forming partnerships with established entities such as ChaCha.com, because this successfully meets users' needs in offering a free and convenient 24/7 service to a large established user base.

SMS is a key aspect of the future of library services. Its successful addition to reference services brings the expertise of librarians to our patrons' mobile phones and ensures the continued relevance of libraries in a mobile world.

27

Say what? An SMS transcript analysis at New York University

Alexa Pearce, Scott Collard and Kara Whatley

Introduction

In this chapter we describe the SMS (Short Message Service) reference service at the New York University Libraries, highlighting the evolution of the service during its first year of operation. We examine in detail a selection of transcripts generated during this period, in order to understand trends both in the nature of users' interactions with SMS reference and in librarians' approaches to staffing the service. The transcript analysis provided an opportunity for us to test common hypotheses about user behaviour, including some of the assumptions that characterized our own initial planning. We gave special attention to three assumptions in particular. The first, that our patrons would use SMS to ask questions that were short and simple, primarily directional in nature. We based this assumption largely on other libraries' published descriptions of their SMS services, in which they reported receiving mostly 'short answer' questions (see for example Hill et al., 2007). The second, that our patrons would want to keep the transaction to as few texts as possible, due to messaging costs. The third, that patrons would likely be texting us from remote locations. As we conducted our analysis, we studied transcripts in the context of these assumptions in an effort to measure actual transactions against many of our expectations and anecdotal observations.

We used a modified question categorization scheme – now a standard tool in the analysis of virtual reference activity (De Groote et al. (2005), Diamond and Pease (2001), Kibbee et al. (2002), Marstellar and Mizzy (2003)) – that enabled us to test many of these hypotheses. The analysis helped us to understand how the service changed, grew and improved over time as librarians become more adept; this prompted us to think about some of the broader implications engendered in offering SMS reference, and in so doing to formulate a better contextual understanding of its role within our overall service infrastructure.

Background

New York University (NYU) is an ARL (Association of Research Libraries) Research One institution situated in the heart of Greenwich Village in New York City. It is the largest private university in the world, with a total enrolment of around 50,000 students, about evenly split between undergraduates and graduates. It has expansive programmes in arts and sciences as well as a number of professional schools offering programmes in education, public policy, social work, business and law. Because of the urban nature of NYU, affiliates of the institution are a mix between on-campus users, off-campus users in private housing and commuters who come from varying distances to attend. There is also a large and growing global representation, with users at sites in eight cities around the world.

This heterogeneous body necessitates a certain flexibility in the provision of services, and NYU has attempted in recent years to expand and enhance its reference provision accordingly. We have tended to staff physical reference services over long hours, in order to accommodate the needs of users with commutes, outside jobs and other time-factor and lifestyle considerations that seem to go along with dwelling in New York City. Our virtual reference portfolio has tried to keep pace with these needs as well, and we currently offer thriving e-mail and IM (instant messaging) services until midnight on most nights of the week. We have observed significant growth in the use of these services during the last few years, indicating to us that our attempts to be user-centred in our approach to service availability have had a welcome effect.

We also try to extend a philosophy of user-centredness to the multiplicity of options that we provide for users to contact us. In addition to offering the multi-modal reference model described above, we also strive to create reference 'front doors' wherever our users are. For instance, in the university-wide information portal, NYUHome, and in our Learning Management System, Blackboard, users find links to Ask-A-Librarian services. We have also incorporated comment and feedback mechanisms throughout our various web interfaces that allow users to comment on particular aspects of our resources and services via Ask-A-Librarian channels (for instance, how our new catalogue interface is working for them, or how they can get help via our Virtual Business Library). Finally, we are expanding the use of IM widgets throughout our sites, starting with our Libguides implementation and moving to other pages from there. Accordingly, our SMS service is the next logical extension of this expansion effort, since it enables us to place our service footprint directly onto our users' mobile devices, providing them with yet another option for contacting us and using our expertise.

Service model and methods

We began our pilot in spring 2008, following several months of research into options for two-way SMS communication. We ultimately decided that there was value in using a single mobile device, as opposed to a complex and expensive software-based system. We investigated the options for devices and carriers before deciding to purchase a Blackberry Pearl. We opened a wireless contract with a carrier who offered a corporate discount for the university, resulting in a total cost-of-entry for one year of service that came to around $1450. Our service model for the pilot was quite simple and entailed assigning the phone to individuals, rather than linking it to an existing reference location, such as one of our desks. We maintained a core group of six to eight librarians who staffed the service for day-long shifts, arranging phone transfers among themselves or retrieving the phone from its central storage location. One person at a time would field all incoming SMS messages during a day, usually between the hours 10:00 a.m. to 6:00 p.m., although some staff also opted to reply to texts received at other times. Official posted hours were for weekdays only, but staff who manned the reference desk at weekends often monitored the phone during those periods as well. Users were able to access the service by texting our dedicated ten-digit number, in the same manner that they would text anyone else. No additional steps or codes were necessary. Initially, our advertising was limited to placing the phone number on our Ask-A-Librarian webpage. Later, we also placed posters throughout the library. Some librarians mentioned the service in classes, tours and other informational events.

Transaction data collection

During our first year of service we kept a log of all SMS transactions. This log is maintained in a password-protected weblog, to which we were able to forward each message directly from the Blackberry. We cleared all identifying information, including phone numbers, from each transaction before saving it. Finally, each transaction was coded, as a comment on its blog entry, for the time that elapsed between receipt of the incoming text and when NYU Library responded to it; the total time of the transaction from start to finish; the total number of messages in the transaction; and the total number of messages received and sent. When we began our transcript analysis, we categorized the transactions along the standard reference/directional axis, as defined by the Association of Research Libraries guidelines. We also noted transactions where a patron was clearly in the library when using the service, and when a patron texted a 'thank you' or similar message to end the transaction as a proxy for unlimited texting. Finally, we spent time analysing the transcripts on a qualitative level to develop our understanding of how the service was being used and staffed.

Results

Transcript analysis

We analysed 693 transcripts dating from April 2008 to April 2009. Many of these transcripts were of a directional nature and contained brief exchanges of two or three messages. These exchanges were the most representative of what we expected the service to look like. In one transcript a user texted to ask what time the library closed on a particular night and received an answer six minutes later in which the librarian said:

> Lower levels 1&2 are open 24 hours, stacks floors are open till 1am, circulation desk open till 10:45. More info @ http://library.nyu.edu/about/hours/spring.html.

The librarian's response is both comprehensive and concise, providing an answer to the immediate question, as well as a link to more complete information online. The entire transaction was contained within a total of two texts and did not include any greetings, expressions of thanks or other conversational elements. Neither the librarian nor the user requested any clarification or indicated a need for follow-up assistance.

Typifying reference service by text

In our transcripts we found that short, directional queries were frequent, but not generally representative of the texting service. As the service has developed over time, it has become increasingly less likely for our exchanges to lack any greeting or salutation. Librarians providing the service have largely dispensed with an initial concern for brevity, even when answering directional questions. In another directional example, a user asked if the library had any computers with a particular type of software, whether there was access to a shared storage drive, and whether or not colour printing was available. The user began the first message with a friendly 'Hi!' and the librarian who responded maintained a pleasant, conversational tone throughout the exchange, which lasted just over two hours and spanned a total of eight texts. The librarian's initial response was two texts in length and was markedly more conversational in nature than the directional example given above. The librarian began with a greeting in the form of 'Hey there', followed by an apology for a slightly delayed reply (close to two hours, but the text was both received and replied to after official closing hours). In addition to a basic answer about using the software in question in the library, the librarian went on to explain how the same resource was also available to individuals:

> Also, did you know that you can get a free 1 year subscription for GIS just by asking the folks who work in the Data Service Studio for a disc?

Had the librarian chosen to omit her greeting, tighten her language, or answer the question as it was asked, without providing additional relevant information, she could have given a shorter response, most likely contained within one text message. The longer, friendlier response that was actually given received a satisfactory response from the user, who replied with 'thanks!' even before the librarian had answered each of the original questions. Accordingly, the librarian texted back to ask, 'Did that answer your question?' and also advised the user on possible places to find more information for one of the questions that she was not able to answer readily, regarding the shared storage drive. The user texted back with a second thank-you message, this time making an explicit comment that the librarian had been thorough and had taken the time to follow up.

The length and tone of this transcript is significant for two reasons. First, the user is demonstrably appreciative that the librarian has been friendly and thorough. This kind of user feedback has enabled us to become increasingly comfortable with answers that span multiple texts – positive responses from users have reinforced our sense that we do not need to be as cautious or conservative with our character space as we originally expected to be. A conversation that takes place via text message will, by nature, be characterized by some degree of brevity, but we have learned from our users that this need not be extreme. Second, in addition to maintaining a conversational tone, the librarian in the above example has used standard reference interview techniques to prompt the user for more information.

Reference service by text is no different

As subsequent examples will demonstrate, reference questions tend to develop the same way in text messages that they do in other venues, which is to say, in response to questions or prompts from the librarian. This alone is reason enough for librarians to ask all necessary questions of their users and not to abbreviate their exchanges, even while communicating via a seemingly abbreviated format.

In one such transcript, a user sends a text asking how to look up a music score in the library's catalogue. Although the librarian is able to give an accurate answer in one text, she responds with the following two messages:

Librarian message 1: Hi, you can limit a search in Bobcat [library catalogue] to scores only. www.bobcat.nyu.edu Format limit is in 1st drop-down menu.

Librarian message 2: Let me know if you'd like help searching. You can send me info about the score & I'll take a look.

The user then responds with details about a 19th-century opera, including time and place of performance. The librarian proceeds to find the score that the user needs, so first tells the user where to find it in the library. In a second text, the librarian also explains exactly how she searched, specifying the terms and mentioning that she omitted any search operators, i.e. 'and'. Finally, she sends a link to the catalogue record and uses two additional text messages to provide some level of instructional context, linking the user to the desired resource.

The example described above occurred over half an hour and spans 11 texts, including a 'thank you' text by the user. Exchanges of this length were common in the transcripts that we looked at, indicating that reference exchanges via text frequently entail back-and-forth conversation, rather than immediate resolution. Not only is it unnecessary to fit an answer into 160 characters, but it tends to work out better for the user if the librarian does not try to do so. We can and should take the time to ask our users relevant questions about their research and address all of their information needs. As in any other reference venue, we do not necessarily get the user's full story unless/until we ask for it.

In some cases, users' tendency to 'return' or to continue texting conversations strongly resembles an in-person interaction in which a user comes back to a reference desk. For example, in one of the longer exchanges that we looked at a total of 16 messages were exchanged over the course of more than three hours. Seven of the messages came from the user and nine were sent by the librarian. The initial text was a statement by the user that he or she needed help 'finding some articles', to which the librarian responded with a question about whether the user was searching on a topic or trying to locate specific citations. The user explained the research topic and class assignment in three subsequent texts, also explaining that he or she did not 'know how to organize this into a search'. Along with database recommendations and relevant web links, the librarian provided very basic guidance for searching and mentioned the possibility of e-mailing the user with 'more specific search strategies', if desired. The user did not respond to the e-mail offer and instead continued to ask, via text, for help in narrowing the search. The user's final message asked specifically if it would be all right to 'message again if necessary'. The librarian then mentioned once more that e-mail was also an option, identifying the topic as one that might be better suited for a longer form of communication, while the user maintained a clear preference for text, regardless of conversation length or complexity.

From these and other transcripts, we have learned that users' preferred modes of communication do not necessarily line up according to predictable models, nor do they necessarily adhere to our expectations.

Text within the library

Finally, we noted several exchanges among the transcripts in which users texted us from within the library. In many cases these messages came from users looking for particular call number locations. In one example the user and the librarian exchanged a total of eight texts about the location of a book, spanning 20 minutes and culminating in an arrangement to actually meet. The user explained that he/she had already checked the books 'waiting to be reshelved', and couldn't find the item needed. The librarian was able to find the book and, after letting the user know this, made arrangements to meet, providing location and identifying information in two texts: 'I am on 2nd floor, in the center, near elevators. Wearing pink turtleneck.'

The initial text came at a time of day when multiple service desks were open, along with our e-mail and IM services. Despite being in close proximity to more than one staffed service desk, the user chose to text a question of an urgent nature. Later in the same day, the user made a decision to text again, even after once stopping by the reference desk.

We see from our analysis of these transcripts that users will text us with all levels and categories of questions and will do so from anywhere. Our decision to respect the user's discretion in selecting an appropriate mode of communication is not an inhibiting factor in the delivery of relevant and comprehensive information.

Discussion

SMS is asynchronous service

When we first offered our SMS reference service, we assumed that patrons would have the expectation of instantaneous answers and we were very careful to post the hours of the pilot service. However, we have found that time is not the major factor in the success or failure of an SMS transaction. Many patrons texted us during our off hours, and when we responded – sometimes hours later or even the next day – they picked up the reference conversation, clarifying information they had received and asking additional questions. We also found that many patrons treat the SMS service just as they would a return visit to the reference desk to talk to the same librarian – they returned to the same conversation, sometimes hours later, for additional assistance.

There is no need to keep answers artificially brief

Initially, we were concerned with keeping our answers to one text message. We did not want to bombard patrons with too much information and we did not want to run up their texting bills. However, our patrons quickly showed us, through their own texting behaviours, that they were not concerned with the number of texts we sent them. With that knowledge, we began treating SMS reference as we would any other type of reference question, giving openings for the patron to ask additional questions and providing follow-up and referral information as appropriate.

SMS happens at point of need – no matter where that is

During the year that we have been offering SMS reference, we have had the chance to talk with a number of librarians about the service and one of the questions that comes up is whether our patrons use the service inside the library building. The answer is a resounding 'Yes!' Our patrons use the service in the stacks when they need help finding a book; they use it from the study tables when a question comes up in their work; they use it in the computer labs when searching the catalogue. They use this service not because it is the only mode available for asking their question, but because it is their preferred mode of communication, no matter where they are. This point is key for the future development of SMS reference: we must realize that it is a preferred mode of communication for many and treat it with the same level of professionalism and quality that we would our other reference transactions.

Practice makes perfect in SMS reference

One of the most interesting things that our SMS transcript analysis showed us is how librarians' approach to SMS reference evolved during the first year. Looking at our transcripts, we could clearly see how the service changed as librarians became more comfortable with SMS. They stopped referring patrons to other modes of reference and they began conducting fuller reference interviews and giving fuller answers to questions, even if it meant sending multiple texts. In short, they treated each SMS query as they would any other reference transaction, and our patrons got better service because of it.

From our overall experiences with the service and our in-depth transcript analysis we have discovered that SMS reference is not so different from any other mode of reference. That is perhaps the biggest lesson to be learned: the evolution of our reference service may involve new technology, but our core skills as reference librarians transfer very nicely, no matter what the mode of communication. The same rules of good practice and good service apply, and each patron and their question must be dealt with on an individual level, with the depth and time desired by the patron.

Conclusion

With our analysis of one year's worth of SMS transcripts, we feel that our SMS reference service demonstrates that the common assumptions we, as librarians, make about text reference are not necessarily borne out under close examination. If the SMS reference service requires us to learn new skills, they are less intellectual in nature than they are mechanical. Learning to provide users with attentive, effective and thorough responses is not unique to conversations that take place via text. Above all, we have learned from our transcripts that our ability to do so is not – and should not be – hampered by the medium.

References

De Groote, S. L., Dorsch, J. L. and Collard, S. (2005) Quantifying Cooperation: collaborative digital reference service in the large academic library, *College & Research Libraries*, 66 (5), 436–54.

Diamond, W. and Pease, B. (2001) Digital Reference: a case study of question types in an academic library, *Reference Services Review*, 29 (3), 210–18.

Hill, J., Hill, C. and Sherman, D. (2007) Text Messaging in an Academic Library: integrating SMS into digital reference, *The Reference Librarian*, 47 (1), 17–29.

Kibbee, J., Ward, D. and Ma, W. (2002) Virtual Service, Real Data: results of a pilot study, *Reference Services Review*, 30 (1), 25–36.

Marsteller, M. R., and Mizzy, D. (2003) Exploring the Synchronous Digital Reference Interaction for Query Types, Question Negotiation, and Patron Response, *Internet Reference Services Quarterly*, 8 (1/2), 149–65.

Conclusion

Gill Needham and Nicky Whitsed

The book following the First International M-Libraries Conference was published in 2008. Two years later, this collection of papers from the second conference indicates the extent to which mobile service development and delivery has taken hold in libraries and information services around the world. What are the trends that underlie this development?

As indicated in many of the papers (chapters 1, 2, 20), mobile devices are increasingly ubiquitous around the world, and both libraries and the publishing industry are therefore considerably more aware of their importance and potential. This increased activity is reflected in the fact that three times as many papers were submitted for the Second International M-Libraries Conference in 2009 as for the first conference in 2007. Mobile delivery of library services has evolved from the largely experimental to the mainstream in many areas.

Although there is ongoing technical innovation, the main emphasis has shifted to implementation and service delivery. Libraries are working to capture the potential of mobile delivery to enhance the service experience they are offering to their users. This would appear to be particularly relevant in an environment where the relevance of library services is being questioned.

In the more prosperous parts of the world, the most sophisticated devices, particularly the iPhone and Android, are gradually becoming affordable. This has enabled developers to be increasingly ambitious and creative (Chapter 11). In contrast, we see some of the most creative implementation plans in the developing world, where telephone companies are helping to make access affordable (chapters 13 and 24) and mobile delivery is becoming mainstream for official information services. These developments will make it possible for everyone to have a virtual library in their pocket.

In the higher education sector, universities are beginning to take an integrated approach to mobile delivery. Examples include the suite of mobile services launched at Ryerson University and spearheaded by the library (Chapter 19) and the podcasting project at the University of British Columbia (Chapter 17).

Most important of all for the future of mobile library services is the development of an evidence base. The surveys and feasibility studies reported in the final section of the book will be invaluable in supporting those who are seeking resources and engagement within their institutions to plan and implement their own services. It will be critical to continue to share the results and experience of studies, however small, to benefit the whole community.

In summary, the volume demonstrates that this emerging field of 'm-libraries' is beginning to mature. There are, however, a number of intriguing questions to consider; for example:

- To what extent will mobile services become mainstream?
- Which devices will be favoured by the majority of library users?
- Will e-book readers have a significant role?
- Will there be a new digital divide between smart and 'unsmart' phones?
- Will publishers engage with delivery to mobile devices?
- What will be the role of librarians in m-libraries?

We look forward to the Third International M-Libraries Conference in 2011, by which time some of these questions will have been addressed.

Index

Making the Most of RFID in Libraries

Martin Palmer

The book is highly recommended for all levels of library and information personnel and for library students and faculty alike...this is a compulsory read and a book I would strongly recommend.

LIBRARY MANAGEMENT

Radio Frequency Identification (RFID) has had a rapid impact on the library world. Its advantage over other technologies used in libraries is usually seen to be its ability to combine the functions of the barcode and the security tag, but with the added advantages of being able to read multiple items seemingly simultaneously without need of line of sight. The customer-friendly self-service that this combination of features makes possible is at the heart of the attraction of RFID for most libraries. This practical and straightforward book is designed to help library managers decide whether RFID has anything to offer them and – if so – how to make the most of the benefits while coping with the challenges inherent in this rapidly developing technology. It also offers many further sources of information to follow up.

Applicable to all types of libraries, its contents include:

- RFID, libraries and the wider world
- RFID in libraries: the background and the basics
- RFID, library applications and the library management system
- standards and interoperability
- privacy
- RFID and health and safety
- RFID and library design
- building a business case for RFID in libraries, and requesting proposals
- staffing: savings, redeployment or something else?
- buying a system: evaluating the offers
- installing RFID: project management
- making the most of RFID: a case study
- RFID, libraries and the future.

RFID has the potential to revolutionize many aspects of library service delivery. Written by an expert in the field, this book is a very worthwhile investment for all those library professionals considering converting to RFID for their libraries, as well as those who are implementing it already.

2009; 176pp; hardback; 978-1-85604-634-3; £44.95

Access, Delivery, Performance
The future of libraries without walls
Edited by Jillian R Griffiths and Jenny Craven

As a conclusion, I would recommend this book not only as a worthy tribute to a prominent figure, but also one that may serve as excellent teaching material in some modern library and information science courses and also as a useful text for professional librarians providing examples of best practice, introducing useful technological tools for library management and work, and depicting some interesting cases from practice in public and academic libraries.

INFORMATION RESEARCH

...this work is a valuable step towards understanding the changes occurring in the field of librarianship as it is subjected to various external pressures and unpredictable user expectations.

AUSTRALIAN ACADEMIC RESEARCH LIBRARIES

This book celebrates and acknowledges the contribution Professor Peter Brophy has made over a career spanning 37 years to the field of library and information studies. Whilst reflecting on his work, it is forward looking and challenging, and offers strategies for the future direction of library and information services in the virtual era.

Following an introduction and tribute to Peter on his retirement, the text is contributed by an international team of acknowledged leaders in their fields, and focuses on four key themes that have preoccupied Peter during his career and that remain of pre-eminent importance for the future of the profession:

- libraries, learning and distance learning
- widening access to information
- changing directions of information delivery
- performance, quality and leadership.

The book concludes with a comprehensive bibliography of Peter's work. This timely book addresses issues and concerns transferable across different areas of the information sector, including academic, public and special libraries, and will be stimulating reading for anyone working, studying, or teaching within the profession.

2008; 256pp; hardback; 978-1-85604-647-3; £44.95

Library Mashups
Exploring new ways to deliver library data
Nicole C Engard

As web users become more savvy and demanding, libraries are looking for new ways to allow user participation. This unique book is geared to help any library keep its website dynamically and collaboratively up-to-date, increase user participation, and provide exemplary web-based service through the power of mashups. Mashups – web applications that combine freely available data from various sources to create something new – can be one very powerful way to meet expectations and provide exemplary web-based service.

This forward-thinking book, with contributions from a team of international experts in the field, brings together definitions, summaries, tools, techniques and real life applications of mashups in libraries. Examples range from ways to allow those without programming skills to make simple website updates to modifying the library OPAC, to using popular sites like Flickr, Yahoo!, LibraryThing, Google Maps and Delicious to share and combine digital content. Key areas covered include:

- what are mashups?
- mashing up library websites
- mashing up catalogue data
- maps, pictures, and videos
- adding value to your services.

A companion website at www.mashups.web2learning.net features an A–Z listing of websites with definitions and examples of mashups, which will be constantly maintained to keep this text completely up to date.

This timely and valuable guide is essential reading for all libraries and librarians seeking a dynamic, interactive web presence. Whether you are a 'newbie' beginner or a veteran programmer, this book is sure to include something that will inspire you and make you think differently about the services your library currently offers.

2009; 312pp; paperback; 978-1-85604-703-6; £29.95

Supporting Research Students

Barbara Allan

The importance of supporting the needs of research students has recently risen higher up the academic agenda around the world. Numbers of postgraduate students have expanded, and the traditional PhD has now been joined by a new range of doctoral qualifications including professional doctorates such as the Doctor in Business Administration (DBA). These developments have led to a more diverse student body which now includes senior professional practitioners.

This shift has seen an acknowledgement that support services within universities must cater more for the needs of research students. While the library and information profession is a graduate one, a relatively small number of LIS professionals have a research degree. This means that, though they are likely to have experience of carrying out smaller scale research projects, they will not have experienced and internalized the distinct learning processes involved in gaining a doctorate.

This timely book offers guidance to enable them to support the specialist needs of research students effectively. Individual chapters are designed to be read and worked through in any order. The key areas covered are:

- research and the research process
- the research student's experience
- research skills training
- supporting research students in academic libraries and information services
- virtual graduate schools
- introduction to research communities
- professional development.

This is an essential text for all library and information professionals in higher education institutions globally that cater for the needs of research students. It will also be valuable reading for LIS students.

2009; 208pp; paperback; 978-1-85604-685-5; £44.95